STUDENT MATHEMATICAL LIBRARY
Volume 42

Invitation to Ergodic Theory

C. E. Silva

AMERICAN MATHEMATICAL SOCIETY
Providence, Rhode Island

Editorial Board

Gerald B. Folland
Robin Forman (Chair)

Brad G. Osgood
Michael Starbird

2000 *Mathematics Subject Classification.* Primary 37A05, 37A25, 37A30, 37A40, 37B10, 28A05, 28A12, 28A20, 28D05, 54E52.

For additional information and updates on this book, visit
www.ams.org/bookpages/STML-42

Library of Congress Cataloging-in-Publication Data
Silva, César Ernesto [date]
  Invitation to ergodic theory / C.E. Silva.
      p. cm. — (Student mathematical library, ISSN 1520-9121 ; v. 42)
  Includes bibliographical references and index.
  ISBN 978-0-8218-4420-5 (alk. paper)
  1. Ergodic theory. I. Title.

QA313.S555 2007
515′.48—dc22

2007060840

Copying and reprinting. Individual readers of this publication, and nonprofit libraries acting for them, are permitted to make fair use of the material, such as to copy select pages for use in teaching or research. Permission is granted to quote brief passages from this publication in reviews, provided the customary acknowledgment of the source is given.

Republication, systematic copying, or multiple reproduction of any material in this publication is permitted only under license from the American Mathematical Society. Requests for permission to reuse portions of AMS publication content are handled by the Copyright Clearance Center. For more information, please visit www.ams.org/publications/pubpermissions.

Send requests for translation rights and licensed reprints to reprint-permission @ams.org.

© 2008 by the American Mathematical Society. All rights reserved.
Reprinted with corrections, 2018.
The American Mathematical Society retains all rights
except those granted to the United States Government.
Printed in the United States of America.

∞ The paper used in this book is acid-free and falls within the guidelines
established to ensure permanence and durability.
Visit the AMS home page at https://www.ams.org/
10 9 8 7 6 5 4 3 2   13 12 11 10 09 08

Contents

Preface		vii
Chapter 1.	Introduction	1
Chapter 2.	Lebesgue Measure	5
§2.1.	Lebesgue Outer Measure	5
§2.2.	The Cantor Set and Null Sets	10
§2.3.	Lebesgue Measurable Sets	17
§2.4.	Countable Additivity	23
§2.5.	Sigma-Algebras and Measure Spaces	26
§2.6.	The Borel Sigma-Algebra	34
§2.7.	Approximation with Semi-rings	38
§2.8.	Measures from Outer Measures	47
Chapter 3.	Recurrence and Ergodicity	59
§3.1.	An Example: The Baker's Transformation	60
§3.2.	Rotation Transformations	67
§3.3.	The Doubling Map: A Bernoulli Noninvertible Transformation	75
§3.4.	Measure-Preserving Transformations	83
§3.5.	Recurrence	86

§3.6.	Almost Everywhere and Invariant Sets	90
§3.7.	Ergodic Transformations	95
§3.8.	The Dyadic Odometer	102
§3.9.	Infinite Measure-Preserving Transformations	109
§3.10.	Factors and Isomorphism	115
§3.11.	The Induced Transformation	120
§3.12.	Symbolic Spaces	123
§3.13.	Symbolic Systems	127

Chapter 4.	The Lebesgue Integral	131
§4.1.	The Riemann Integral	131
§4.2.	Measurable Functions	134
§4.3.	The Lebesgue Integral of Simple Functions	141
§4.4.	The Lebesgue Integral of Nonnegative Functions	145
§4.5.	Application: The Gauss Transformation	151
§4.6.	Lebesgue Integrable Functions	155
§4.7.	The Lebesgue Spaces: L^1, L^2 and L^∞	159
§4.8.	Eigenvalues	166
§4.9.	Product Measure	170

Chapter 5.	The Ergodic Theorem	175
§5.1.	The Birkhoff Ergodic Theorem	176
§5.2.	Normal Numbers	188
§5.3.	Weyl Equidistribution	191
§5.4.	The Mean Ergodic Theorem	192

Chapter 6.	Mixing Notions	201
§6.1.	Introduction	201
§6.2.	Weak Mixing	205
§6.3.	Approximation	209
§6.4.	Characterizations of Weak Mixing	214
§6.5.	Chacón's Transformation	218
§6.6.	Mixing	226

Contents

§6.7.	Rigidity and Mild Mixing	227
§6.8.	When Approximation Fails	231

Appendix A. Set Notation and the Completeness of $\mathbb{R}$ — 235

Appendix B. Topology of $\mathbb{R}$ and Metric Spaces — 241

Bibliographical Notes — 251

Bibliography — 255

Index — 259

Preface

This book provides an introduction to the growing field of ergodic theory, also known as measurable dynamics. It covers topics such as recurrence, ergodicity, the ergodic theorem and mixing. It is aimed at students who have completed a basic course in undergraduate real analysis covering topics such as basic compactness properties and open and closed sets in the real line. Measure theory is not assumed and is developed as needed. Readers less familiar with these topics will find a discussion of the relevant material from real analysis in the appendices.

I have used early versions of this book in courses that are designed as capstone courses for the mathematics major, including students with a variety of interests and backgrounds. The study of measurable dynamics can be used to reinforce and apply the student's knowledge of measure theory and real analysis while introducing some beautiful mathematics of relatively recent vintage. Measure theory is developed as needed and applied to study notions in dynamics. While it has less emphasis, some metric space topology, including the Baire category theorem, is presented and applied to topological dynamics. Several examples are developed in detail to illustrate concepts from measurable and topological dynamics.

This book can be used as a special-topics course for upper-level mathematics students. It can also be used as a short introduction to Lebesgue measure and integration, as an introduction to ergodic theory, and for independent study. The Bibliographical Notes provide some guidelines for further reading.

An introductory course could start with a short review of the topology on the real line and basic properties of metric spaces as covered in Appendix B or with the construction of Lebesgue measure on the real line in Chapter 2. The reader who wants to get to ergodic theory quickly needs to cover only Sections 2.1 through 2.4 and could then start with Chapter 3, perhaps omitting Sections 3.10 through 3.12. Topological dynamics is closely related to measurable dynamics, and the book introduces some topics from topological dynamics. This is not necessary for the main development of the book, and the reader has the option of omitting the topological dynamics topics or of using them to learn some metric space topology and some elegant ideas from topological dynamics. A more advanced course could cover Chapters 2 and 3 in more detail. Some of the measure theory notions that are covered include the Carathéodory extension theorem, product measures and L^p spaces. Lebesgue integration is introduced in Chapter 4, and some of these notions are used to study the eigenvalues of measure-preserving transformations. The chapter on the ergodic theorem, in addition to being of intrinsic interest, provides a beautiful example for applications of various theorems of Lebesgue integration. The final chapter on mixing uses ideas from all the other chapters.

The book contains both simple exercises, called questions, designed to test the reader's immediate grasp of the new material, and more challenging exercises at the end of each section. Harder exercises are marked with a star ($\star$). Partial solutions and hints for some of the exercises will be available at the book's webpage listed on the back cover. Some sections also contain open questions designed to suggest to the reader some avenues of research. The bibliography is not intended to be exhaustive; it is there to provide suggestions for additional reading and to acknowledge the sources I have used.

I am indebted to many people who through their conversations and writings have taught me measure theory and ergodic theory. I first learned analysis from César Carranza. I was introduced to

Preface ix

ergodic theory by Dorothy Maharam and later was influenced by Shizuo Kakutani and John Oxtoby. I have also learned much from all my coauthors and the students I have supervised in research and in courses. I am indebted to several anonymous reviewers and readers at various stages of this work who have provided advice and suggestions. In particular I would like to thank my Williams colleagues Ollie Beaver, Ed Burger, Satyan Devadoss, Frank Morgan and Mihai Stoiciu, and my editor, Sergei Gelfand. I also thank Blaire Madore and Karin Reinhold, who used an early version with their students and sent me helpful suggestions.

I have used early versions of this book in courses, tutorials and summer SMALL REU projects and would like to thank the many students who corrected errors, discovered typos and made suggestions, in particular Katherine Acton, Ali Al-Sabah, Amie Bowles, John Bryk, John Chatlos, Tegan Cheslack-Postava, Alexandra Constantin, Jon Crabtree, Michael Daub, Sarah Day, Chris Dodd, Jason Enelow, Lukasz Fidkowski, Thomas Fleming, Artur Fridman, Ilya Grigoriev, Brian Grivna, Kate Gruher, Fred Hines, Sarah Iams, Catalin Iordan, Nate Ince, On Jesakul, Anne Jirapattanakul, Jeff Kaye, Eric Katerman, Brian Katz, Min Kim, James Kingsbery, Dave Kleinschmidt, Thomas Koberda, Ross Kravitz, Gary Lapon, Alex Levin, Amos Lubin, Amy Marinello, Earle McCartney, Abe Menon, Erich Muehlegger, Karl Naden, Nara Narasimhan, Deepam Patel, Ravi Purushotham, Andy Raich, Hyejin Rho, Becky Robinson, Richard Rodriguez, David Roth, Charles Samuels, Brian Simanek, Peter Speh, Anita Spielman, John Spivack, Noah Stein, Joseph Stember, Andrea Stier, Brian Street, Daniel Sussman, Mike Touloumtzis, Paul Vichyanond, Robert Waelder, Kirsten Wickelgren, Alex Wolfe, and Wenhuan Zhao. Thanks are due most especially to Darren Creutz, Daniel Kane, Jennifer James, Kathryn Lindsey, and Anatoly Preygel.

Finally, I would like to dedicate this book to my wife and two daughters and the memory of my parents.

Cesar E. Silva

Chapter 1

Introduction

Dynamics is a discipline with roots that go back at least to the time of Newton and questions about the motion of heavenly bodies. More recently, dynamics has grown into a broad discipline that encompasses differential equations, population biology models and group actions on homogeneous spaces. In the late 1800's, Poincaré initiated a qualitative approach to dynamical problems and, following this approach, various fields have emerged where the phase space of the dynamical system (the space that represents all possible states of a system in motion) is an abstract mathematical space. A dynamical system, in this more abstract approach, consists of a set X and a function or transformation T defined on X and with values in X. In ergodic theory, also known as measurable dynamics, X and T are given a measurable structure. We will occasionally deal with questions from topological dynamics, where X is a topological space, usually a compact metric space, and T a continuous map or a homeomorphism.

Ergodic theory uses tools of measure theory to study the long-term behavior of abstract dynamical systems. Its origins lie in statistical mechanics with, for example, the study of hard-sphere gases. While this is outside the scope of this book, we describe briefly how a dynamical system is obtained from this setting. To study the time evolution of n particles in space, each particle can be represented by a position vector and a momentum vector, so with six numbers for each particle, a point in $\mathbb{R}^{6n}$ can be thought of as representing the state of the system at a moment in time. The physical laws that govern the motion can be modeled by differential equations and, using a classical theorem of Liouville, it can be shown that there are regions of the phase space, which sometimes are compact sets, where there is a measure that respects the motion (a concept that is called an invariant measure and that is studied in Chapter 3). In this way one obtains a law of motion, in this case a flow, defined on a measure space.

In the system that we just discussed, as in most systems arising in nature, time is continuous. Before describing a reduction to the discrete-time case we introduce some notation. Let x be a point in the phase space and let $T_t(x)$ represent the state of the system after t units of time have elapsed from the moment the system is in state x. We observe two properties of T_t. First, when $t = 0$ the map T_0 should be the identity transformation, and next if $T_{t_1}(x)$ is the state of the system after t_1 units of time have elapsed, the state after an additional t_2 units of time should be the same as the state after $t_1 + t_2$ units of time have elapsed from the initial configuration. So the family of maps $\{T_t\}$ should satisfy the following equations:

$$T_0(x) = x,$$
$$T_{t_1+t_2}(x) = T_{t_1}(T_{t_2}(x)).$$

A family of maps T_t satisfying this "time invariance" property is called a flow; we think of it as modeling the evolution of continuous time in a system. While continuous-time systems are the first systems to arise in a natural way, discrete-time dynamical systems had already been suggested by Poincaré as good approximations of continuous systems, and they are important as they show the qualitative range of behavior (such as ergodicity and mixing) that is present in continuous-time systems. For example, a motion picture is a discrete approximation of the reality it is filming. A discrete-time dynamical system may be obtained from any continuous system by fixing an initial time and considering its integer multiples. Thus, to approximate the dynamics of a continuous system one can fix a "small" time t_0 and analyze the motion of the systems at times that are integer multiples of t_0. More precisely, define a self-map or transformation T of the measure space

1. Introduction

by $T(x) = T_{t_0}(x)$. So, given an initial state x, $T(x)$ represents the state of the system after t_0 units of time. Then $T^2(x) = T(T(x)) = T_{t_0+t_0}(x)$ would represent the state of the system at time $t_0+t_0 = 2t_0$ units of time, and $T^n(x)$ is similarly defined to represent the state of the system after nt_0 units of time when starting at x. Thus, one arrives at discrete-time dynamical systems defined on measure spaces, the subject on which this text focuses.

The next important simplification in ergodic theory reduces the phase space X to an interval in $\mathbb{R}$, or more generally, a Legesgue space. It can be shown that measure-theoretically (i.e., up to a measurable isomorphism) all nonpathological spaces (e.g., complete separable metric spaces with a Borel measure) are isomorphic to a finite or infinite interval in $\mathbb{R}$ (with possibly a countable number of isolated points of positive mass called atoms). Therefore, our first study is that of transformations T on a space X that is usually a finite, or in some cases infinite, interval in $\mathbb{R}$. Chapter 2 develops the theory of Lebesgue measure on the real line, and then the basic concepts of measure theory that are needed for the study of the dynamical properties of recurrence and ergodicity. Some examples of dynamical systems are more naturally described on other measure spaces such as symbolic spaces or subsets of $\mathbb{R}^d$, so we briefly discuss how measure theory on the line generalizes to these spaces.

An important development, that we do not cover in this book, in ergodic theory was the introduction by Kolmogorov of the notion of entropy. Using entropy as developed by Kolmogorov and Sinai, Ornstein proved in 1970 a remarkable theorem classifying Bernoulli automorphisms up to isomorphism. Our development of ergodic theory stops short of the notion of entropy and we refer to the Bibliographical Notes for further references.

While ergodic theory originated in statistical mechanics and initially dealt only with measure-preserving transformations on finite measure spaces, already in the 1930's mathematicians were interested in infinite measure spaces, and later in relaxing the measure-preserving condition and considering what we call nonsingular transformations. We emphasize finite measure-preserving transformations but also consider notions such as recurrence and ergodicity in the context of infinite measure-preserving transformation in Chapter 3.

Equally important as the examples that have come from physics are examples and questions that originated in other areas of mathematics. One of the early results in ergodic theory is the beautiful theorem of Weyl on the equidistribution of numbers, which is studied in Chapter 5. A more recent example, that is only mentioned without proof in this book, is Furstenberg's ergodic theoretic proof, published in 1977, of the celebrated theorem of Szemerédi on arithmetic progressions. This theorem states that any set of integers of positive upper (Banach) density has infinitely many arithmetic progressions of a given length. To prove this theorem Furstenberg associated to each set of integers of positive density a finite measure space and a transformation defined on it. He then proved a remarkable theorem known as the Multiple Recurrence Theorem which, under this correspondence, implies the Szemerédi theorem. The methods in Furstenberg's proof have proven useful in solving other problems. In particular they have played a role in Green and Tao's recent solution of a 300-year old problem on the existence of arithmetic progressions in the primes.

Chapter 2

Lebesgue Measure

This chapter develops the basic notions of measure theory. They are what is needed to introduce the concepts of measure-preserving transformations, recurrence and ergodicity in Chapter 3. We first develop the theory of Lebesgue measure on the real line. As we shall see, all the basic ideas of measure are already present in the construction of Lebesgue measure on the line. We end with a section on the changes that are necessary to extend our construction of Lebesgue measure from the real line to d-dimensional Euclidean space. The reader is referred to the appendices for mathematical notation not defined here and for basic properties of the real numbers that we use.

2.1. Lebesgue Outer Measure

Lebesgue measure in $\mathbb{R}$ generalizes the notion of length. We will see that the notion of length is generalized to a large class of subsets of the line.

The simplest sets for which we have a good notion of length are the intervals, and they form the starting point for our development of Lebesgue measure. The length of an interval I is denoted by $|I|$. In Theorem 2.1.3 we show that, as expected, the Lebesgue measure of an interval is indeed equal to its length.

Sets will be "measured" by approximating them by countable unions of intervals. We will approximate our sets "from above," i.e., we will consider unions of intervals *containing* our sets. The idea is to consider all possible countable collections of intervals covering a given set A and to take the sum of the lengths of the respective intervals. Then to obtain the "measure" we take the infimum of all these numbers. Formally, we define the **Lebesgue outer measure** or simply the **outer measure** of a set A in $\mathbb{R}$ by

(2.1)
$$\lambda^*(A) = \inf\left\{\sum_{j=1}^{\infty} |I_j| : A \subset \bigcup_{j=1}^{\infty} I_j, \text{where } I_j \text{ are bounded intervals}\right\}.$$

The set over which the infimum is taken in (2.1) (i.e., the set of sums of lengths of intervals) is bounded below by 0. Thus, if one such sum in (2.1) is finite, the completeness property of the real numbers implies that $\lambda^*(A)$ is a (finite) nonnegative real number. It may happen that for all intervals covering A, the sum of their lengths is ∞; in this case we write $\lambda^*(A) = \infty$ (we will see that this happens if, for example, $A = \mathbb{R}$). With the understanding that the outer measure may be infinite in some cases, we see that the notion of outer measure is defined for every subset A of $\mathbb{R}$.

It is reasonable to ask what happens if one only takes finite sums in (2.1) instead of infinite sums. This notion is called Jordan content or Peano-Jordan content and it does not yield a countably additive measure.

We are now ready to state some basic properties of the outer measure of a set.

Proposition 2.1.1. *Lebesgue outer measure satisfies the following properties.*

(1) *The intervals I_j in the definition of outer measure may all be assumed to be open.*

(2) *For any constant $\delta > 0$, the intervals I_j in the definition of outer measure may all be assumed to be of length less than δ.*

(3) *For any sets A and B, if $A \subset B$, then $\lambda^*(A) \leq \lambda^*(B)$.*

2.1. Lebesgue Outer Measure

(4) (*Countable Subadditivity*) For any sequence of sets $\{A_j\}$ in $\mathbb{R}$ it is the case that
$$\lambda^*(\bigcup_{j=1}^{\infty} A_j) \leq \sum_{j=1}^{\infty} \lambda^*(A_j).$$

Proof. For part (1) let $\alpha(A)$ denote the outer measure of A when computed using only open bounded intervals in the coverings. Clearly, $\lambda^*(A) \leq \alpha(A)$. Now let $\varepsilon > 0$. For any covering $\{I_j\}$ of A let K_j be an open interval containing I_j such that $|K_j| < |I_j| + \frac{\varepsilon}{2^j}, j \geq 1$. Then
$$\sum_{j=1}^{\infty} |K_j| < \sum_{j=1}^{\infty} |I_j| + \sum_{j=1}^{\infty} \frac{\varepsilon}{2^j} = \sum_{j=1}^{\infty} |I_j| + \varepsilon.$$
Taking the infimum of each side gives $\alpha(A) \leq \lambda^*(A) + \varepsilon$, and as this holds for all ε, $\alpha(A) \leq \lambda^*(A)$.

Part (2) follows from the fact that intervals can be subdivided into subintervals without changing the sum of their lengths.

Part (3) follows directly from the definition.

For (4), first note that if for some $j \geq 1$, $\lambda^*(A_j) = \infty$, then there is nothing to prove. Suppose now that for all $j \geq 1$, $\lambda^*(A_j) < \infty$. Let $\varepsilon > 0$. Applying Lemma A.1.1, for each $j \geq 1$ there exist intervals $\{I_{j,k}\}_{k \geq 1}$ such that
$$A_j \subset \bigcup_{k=1}^{\infty} I_{jk} \quad \text{and} \quad \sum_{k=1}^{\infty} |I_{j,k}| < \lambda^*(A_j) + \frac{\varepsilon}{2^j}.$$
Then
$$A \subset \bigcup_{j=1}^{\infty} \bigcup_{k=1}^{\infty} I_{j,k}$$
and
$$\lambda^*(A) \leq \sum_{j=1}^{\infty} \sum_{k=1}^{\infty} |I_{j,k}| < \sum_{j=1}^{\infty} (\lambda^*(A_j) + \frac{\varepsilon}{2^j})$$
$$= (\sum_{j=1}^{\infty} \lambda^*(A_j)) + \varepsilon.$$
Since this holds for all $\varepsilon > 0$, $\lambda^*(A) \leq \sum_{j=1}^{\infty} \lambda^*(A_j)$. $\square$

We require the following basic lemma about the notion of length.

Lemma 2.1.2. *Let $[a,b]$ be a closed interval and let $\{I_j\}_{j=1}^{K}$ be any finite collection of open intervals such that $[a,b] \subset \bigcup_{j=1}^{K} I_j$. Then*

$$|[a,b]| \leq \sum_{j=1}^{K} |I_j|.$$

Proof. If any I_j is infinite, then $\sum_{j=1}^{K} I_j = \infty$, so we need consider only the case when all the intervals I_j are finite. Write $I_j = (a_j, b_j)$, for $1 \leq j \leq K$. Let j_1 be the smallest integer such that $a \in (a_{j_1}, b_{j_1})$. Let j_2 be the smallest integer such that the previous point $b_{j_1} \in (a_{j_2}, b_{j_2})$. This generates a sequence $j_1, j_2, \ldots$ which must terminate, as the collection of intervals from which it comes is finite. From the construction it must terminate at some integer j_ℓ such that $a_{j_\ell} < b < b_{j_\ell}$. Then

$$b - a < b_{j_\ell} - a_{j_1} \leq \sum_{i=1}^{\ell} |I_{j_i}| \leq \sum_{j=1}^{K} |I_j|.$$

□

The following theorem shows that indeed Lebesgue outer measure generalizes our notion of length.

Theorem 2.1.3. *If I is a (bounded or unbounded) interval, then*

$$\lambda^*(I) = |I|.$$

Proof. Let I be a bounded interval. As I covers itself, we have that $\lambda^*(I) \leq |I|$, so it suffices to show that $|I| \leq \lambda^*(I)$. First assume that $I = [a,b]$, a closed bounded interval. Calculate outer measure using open bounded intervals. Let $\{I_j\}_{j=1}^{\infty}$ be a sequence of open bounded intervals covering I. By Theorem B.1.5, there exists a finite subcollection $\{I_{j_i}\}_{i=1}^{K}$ of these intervals that covers I. By Lemma 2.1.2,

$$b - a \leq \sum_{i=1}^{K} |I_{j_i}| \leq \sum_{j=1}^{\infty} |I_j|.$$

This means that $|I| \leq \lambda^*(I)$.

2.1. Lebesgue Outer Measure

Next consider any bounded interval I. For each $\varepsilon > 0$ choose a closed interval $J_\varepsilon \subset I$ such that $|J_\varepsilon| > |I| - \varepsilon$. Then

$$|I| < |J_\varepsilon| + \varepsilon = \lambda^*(J_\varepsilon) + \varepsilon \leq \lambda^*(I) + \varepsilon.$$

Since this holds for all $\varepsilon > 0$, $|I| \leq \lambda^*(I)$, which is the desired inequality.

Finally, consider an unbounded interval I. Then for any integer $k > 0$ there is a bounded interval $J \subset I$ with $|J| = k$. Therefore, $\lambda^*(I) = \infty$. □

In closing, define a set N to be a **null set** if its outer measure is zero, i.e., $\lambda^*(N) = 0$. Proposition 2.1.1(4) then implies that countable sets are null sets. An interesting consequence of this and Theorem 2.1.3 is another proof that intervals are not countable. In Section 2.2 we see that the Cantor set provides an example of an uncountable set that is a null set. Cantor sets of positive measure (see Exercise 2.4.5) provide examples of sets that contain no intervals but are not null.

Exercises

(1) Show that there is no greater generality in the definition of outer measure if the intervals are not restricted to being bounded.

(2) a) Show that the intervals in the definition of outer measure may be assumed to be closed. b) Show that for any δ, the intervals in the definition of outer measure may be assumed to be open and of length less than δ. c) Show that if A is contained in an interval K, then we can assume that all intervals I_j in the cover are contained in K.

(3) A **dyadic interval** or 2-adic interval is an interval of the form

$$[k/2^j, (k+1)/2^j)$$

for some integers k and $j \geq 0$. (It is convenient to take them left-closed and right-open.) Show that the intervals I_j in the definition of outer measure may all be assumed to be dyadic intervals.

(4) For any set $A \subset \mathbb{R}$ and number t, define $A+t = \{a+t : a \in A\}$, the translation of A by t. Show that $\lambda^*(A+t) = \lambda^*(A)$.

(5) Show that if N is a null set, then for any set A, $\lambda^*(A \cup N) = \lambda^*(A)$.

(6) Show that if a set is bounded, then its outer measure is finite. Is the converse true?

(7) Show that the union of countably many null sets is a null set.

(8) For any $t \in \mathbb{R}$ define $tA = \{ta : a \in A\}$. Show that $\lambda^*(tA) = |t|\lambda^*(A)$.

(9) Generalize Exercise 3 to the case of q-adic intervals, $q > 2$ (similar to a 2-adic interval but with 2 replaced by q).

∗ (10) In the definition of outer measure of a set A, replace countable intervals covering A by finite intervals covering A and call it *outer content*. Determine which of the properties that we have shown for outer measure still hold for outer content.

2.2. The Cantor Set and Null Sets

This section introduces the **Cantor middle-thirds set** K and its basic properties. The Cantor set is a remarkable set that plays a crucial role as a source of examples and counterexamples in analysis and dynamics.

The set will be defined inductively. Let
$$F = [0,1] \text{ and } G_0 = (\frac{1}{3}, \frac{2}{3}).$$
We say that G_0 is the *open middle-third* of F. Then set
$$F_1 = F \setminus G_0.$$
F_1 consists of 2^1 closed intervals in F, each of length $\frac{1}{3}$ and denoted by $F[0]$ and $F[1]$. Let $G_1 = (\frac{1}{9}, \frac{2}{9}) \cup (\frac{7}{9}, \frac{8}{9})$, the union of the middle-thirds of each of the subintervals of F_1. Then set
$$F_2 = F_1 \setminus G_1.$$

2.2. The Cantor Set and Null Sets

F_2 consists of 2^2 closed subintervals of F_1, each of length $\frac{1}{3^2}$ and denoted, in order from left to right, by $F[00], F[01], F[10], F[11]$. Figure 2.1 shows the first few steps in the construction.

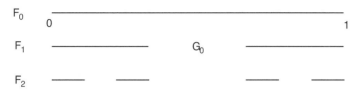

Figure 2.1. First steps in the construction of the Cantor set

Now suppose that F_{n-1} has been defined and consists of 2^{n-1} closed subintervals, each of length $\frac{1}{3^{n-1}}$ and denoted by (ordered from left to right)
$$F[0\cdots 0], F[0\cdots 1], \ldots, F[1\cdots 1].$$
Let G_{n-1} be the union of the open middle-thirds (each of length $\frac{1}{3^n}$) of each of the 2^{n-1} closed subintervals of F_{n-1}. Then set
$$F_n = F_{n-1} \setminus G_{n-1},$$
a union of 2^n closed subintervals, each of length $\frac{1}{3^n}$. The Cantor set K is defined by
$$K = \bigcap_{n=1}^{\infty} F_n.$$
It follows from this definition that all endpoints of the subintervals in F_n belong to K. It might seem that "most" points of $[0,1]$ have been removed, but in fact, as we shall see in the theorem below, there are uncountably many points that are left in K. The next exercise provides an alternative way to think about K.

Question. Show that the Cantor set is also given by
$$K = F \setminus \left(\bigcup_{n=0}^{\infty} G_n \right).$$
Thus, the Cantor set is the set of points in $[0,1]$ that is obtained after removing all the open intervals comprising the sets G_n.

It is interesting to note the following characterization of a null set; its proof is left to the reader.

2. Lebesgue Measure

Lemma 2.2.1. *A set N is a null set if and only if for any $\varepsilon > 0$ there exists a sequence of intervals I_j such that*

$$N \subset \bigcup_{j=1}^{\infty} I_j \text{ and } \sum_{j=1}^{\infty} |I_j| < \varepsilon,$$

where $|I_j|$ denotes the length of the interval I_j.

The idea is that the set N can be covered by intervals such that the sum of their lengths can be made arbitrarily small. A set consisting of a single point p is clearly a null set, as it is covered by $(p - \varepsilon, p + \varepsilon)$ for all $\varepsilon > 0$. (As intervals are allowed to be points, a simpler proof could be obtained by taking the interval covering $\{p\}$ to be $I = [p,p]$, but we have given a proof that also works with the more restrictive definition requiring the covering intervals to be open, or of positive length. Exercise 3 shows that both definitions are equivalent.) The reader may verify that countable sets are null sets. Surprisingly, there exist uncountable sets that are null sets. An important example of this is provided by the Cantor set.

Theorem 2.2.2. *The Cantor set K is a null set and an uncountable closed subset of $[0, 1]$. Furthermore, K contains no positive length intervals and is perfect, i.e., every point of K is an accumulation point of K.*

Proof. We know that

$$K = \bigcap_{n=1}^{\infty} F_n,$$

where F_n is a union of 2^n closed intervals, each of length $1/3^n$. Given $\varepsilon > 0$ choose n so that $(2/3)^n < \varepsilon$. Then the sum of the lengths of the intervals comprising F_n is less than ε. Therefore K is covered by a finite union of intervals whose total length is less than ε, showing that K is a null set. Also, K is closed as it is an intersection of closed sets.

If I is an interval contained in K, then $I \subset F_n$ for all $n > 0$. As I is an interval, it must be contained in one of the subintervals of F_n, each of length $1/3^n$. So $|I| < 1/3^n$, for all $n > 0$. Therefore K contains no intervals of positive length.

2.2. The Cantor Set and Null Sets

To show that K is uncountable we define the following function. First, represent $x \in [0,1]$ in binary form as

$$x = \sum_{i=1}^{\infty} \frac{x_i}{2^i}, \text{ where } x_i \in \{0,1\}.$$

For example $\frac{1}{3}$ is represented by

$$\frac{0}{2} + \frac{1}{2^2} + \frac{0}{2^3} + \frac{1}{2^4} + \cdots$$

since

$$\sum_{i=1}^{\infty} \frac{1}{2^{2i}} = \sum_{i=1}^{\infty} \frac{1}{4^i} = \frac{1}{3}.$$

This representation is unique if we assume that it does not end in an infinite sequence of 1's. More precisely, let D consist of all the numbers in $[0,1]$ of the form $\frac{k}{2^n}$, for integers $k \geq 0, n > 0$, and let $X_0 = [0,1] \setminus D$. D is countable and consists precisely of the numbers in $[0,1]$ that have more than one representation in binary form. (For example $\frac{1}{2}$ may be written in one way with $x_1 = 1$ and $x_i = 0$ for $i \geq 2$ and in another way with $x_1 = 0$ and $x_i = 1$ for $i \geq 2$.) So each $x \in X_0$ has a unique binary representation. For

$$x = \sum_{i=1}^{\infty} \frac{x_i}{2^i} \in X_0$$

define the map $\phi : X_0 \to K$ by

(2.2) $$\phi(x) \in \bigcap_{n=1}^{\infty} F[x_1 x_2 \cdots x_n].$$

To show that ϕ is a function we need to verify that the intersection in (2.2) consists of a single point. Note that since the sets $F[x_1 x_2 \cdots x_n]$ are closed and are contained in F_n and since

$$\lim_{n \to \infty} \lambda(F_n) = 0,$$

the intersection $\bigcap_{n=1}^{\infty} F[x_1 x_2 \cdots x_n]$ contains a unique point, and this point must be in K. So ϕ is a well-defined function from X_0 into K. To show that K is uncountable it suffices to show that ϕ is one-to-one. (ϕ is also onto K minus the endpoints of $F_n, n \geq 1$, but that is not needed here and is left to the reader as an exercise.) Let $x, y \in X_0$. Then $x = \sum_{i=1}^{\infty} x_i/2^i$ and $y = \sum_{i=1}^{\infty} y_i/2^i$. Suppose $x \neq y$.

Then there exists some $k > 0$ such that $x_k \neq y_k$. This means that the sets $F[x_1 x_2 \cdots x_k]$ and $F[y_1 y_2 \cdots y_k]$ are disjoint, which implies that $\phi(x) \neq \phi(y)$, so ϕ is one-to-one. It follows that the cardinality of K must be at least that of X_0, so K must be uncountable.

To see that K is perfect, for any $x \in K$ and for any $n > 0$ let a_n be any endpoint, different from x, of the subinterval in F_n that contains x. Then $\{a_n\}$ is a sequence of points of K different from x and converging to x (as $|x - a_n| < 1/2^n$). So K is perfect. □

A set is said to be **nowhere dense** if its closure has empty interior, i.e., its closure contains no open sets. This is the same as saying that there is no (nonempty) interval in which the set is dense. Evidently, any finite set in $\mathbb{R}$ is nowhere dense, but infinite sets may also be nowhere dense (such as $\mathbb{Z}$ or $\{1/n\}_{n>0}$). A set is said to be **totally disconnected** if its *connected components* (not defined here) are just points. It can be shown for the case of subsets of $\mathbb{R}$ that a set is totally disconnected if and only if it contains no positive length intervals, and we adopt this as our definition of totally disconnected for a subset of $\mathbb{R}$. Evidently, a closed set in $\mathbb{R}$ is nowhere dense if and only if it is totally disconnected. We have just shown that the Cantor middle-thirds set is a closed, bounded, perfect and totally disconnected subset of $\mathbb{R}$. Any subset of $\mathbb{R}$ satisfying these properties is called a **Cantor set**. The notion of a Cantor set can be defined for more general sets, but one needs the notion of a *homeomorphism* between topological spaces. Then using these notions (not defined in this book) a more general Cantor set can be defined as any topological space that is homeomorphic to the Cantor middle-thirds set.

We have observed that the notion of Lebesgue measure zero coincides with the notion of null set. Using the notion of measure, another way to see that K is a set of measure zero is to show that its complement G in $[0,1]$ has measure 1. This is a simple computation as the sum of the lengths of the intervals comprising G is

$$\lambda(G) = \sum_{n=0}^{\infty} \frac{2^n}{3^{n+1}} = 1.$$

We shall see in Theorem 2.4.1 that if $K \sqcup G = [0,1]$, then the measure of K is $1 - \lambda(G) = 0$. One can intuitively think that K has measure

2.2. The Cantor Set and Null Sets

zero as it is the set that remains in $[0,1]$ after removing a set G of measure 1. This is also the starting point for constructing other Cantor sets that are not null; they are obtained after removing a disjoint countable collection of intervals such that the sum of their lengths add to less than 1 (see Exercise 2.4.(5)).

Define a transformation $T : \mathbb{R} \to \mathbb{R}$ by

$$T(x) = \begin{cases} 3x & \text{if } x \leq \frac{1}{2}; \\ 3 - 3x & \text{if } x > \frac{1}{2}. \end{cases}$$

A set that is interesting to consider in dynamics is the set of points x whose positive orbit under T, i.e., the set $\{T^n(x) : n \geq 0\}$, is bounded. So define Λ to be such that if $x \in \Lambda$, then the positive orbit $\{T^n(x)\}_{n \geq 0}$ is bounded. Note that if $x \in \Lambda$, then $T(x) \in \Lambda$. For the case of the transformation T first observe that if $x < 0$, then $T(x) = 3x < 0$, and by induction $T^n(x) = 3^n x$ for all $n > 0$. It follows that if $x < 0$, or $T^k(x) < 0$ for some k, then $T^n(x) \to -\infty$ as $n \to \infty$. Next observe that if $T^n(x) > 1$ for some $n \geq 0$, then $T^{n+1}(x) = T(T^n(x)) < 0$, so $x \notin \Lambda$. Clearly, $0 \in \Lambda$. Thus Λ is characterized by

(2.3) $\qquad \Lambda = \{x \in [0,1] : T^n(x) \in [0,1] \text{ for all } n \geq 0\}.$

Using (2.3), we now identify Λ with the middle-thirds Cantor set K. For this write each point $x \in K$ in its ternary representation. If $x \in K \cap [0, \frac{1}{2}]$, then

$$x = \sum_{i=2}^{\infty} \frac{x_i}{3^i}.$$

So,

$$T(x) = 3x = \sum_{i=2}^{\infty} \frac{x_i}{3^{i-1}} = \sum_{i=1}^{\infty} \frac{x_{i+1}}{3^i}.$$

This shows that $T(x)$ is in K. In addition it describes the effect of T on the point x: we see that if the digits in the ternary representation of x are $x_1 x_2 \cdots$, then the digits in the ternary representation of $T(x)$ are $x_2 x_3 \cdots$. In other words, T shifts the representation of x to the left. Now if $x \in K \cap [\frac{1}{2}, 1]$, then

$$x = \frac{2}{3} + \sum_{i=2}^{\infty} \frac{x_i}{3^i}.$$

So,

$$T(x) = 3 - 3(\frac{2}{3} + \sum_{i=2}^{\infty} \frac{x_i}{3^i}) = \sum_{i=1}^{\infty} \frac{2 - x_{i+1}}{3^i}.$$

So, $T(x) \in K$. In a similar way one can show that if $x \in [0,1] \setminus K$, then $T^k(x) > 1$ for some $k > 0$, so $x \notin \Lambda$. One concludes that $K = \Lambda$.

Exercises

(1) Prove Lemma 2.2.1.

(2) Let F_n be as in the construction of the Cantor set K. Show that for all n, the endpoints of the closed subintervals in F_n belong to the Cantor set.

(3) Show that in Lemma 2.2.1 the sets I_j may be assumed to be nonempty open intervals.

(4) Modify the construction of the Cantor middle-thirds set in the following way. At the first stage remove a central interval of length $\frac{1}{5}$ and at the nth stage, instead of removing the open middle-thirds, remove the open middle-fifth of each subinterval. Show that in this way you obtain a Cantor set that is a null set.

(5) Recall that every $x \in [0,1)$ can be written in ternary expansion as $x = \sum_{i=1}^{\infty} \frac{x_i}{3^i}$, where $x_i \in \{0,1,2\}$. Show that the Cantor set K is precisely $K = \{\sum_{i=1}^{\infty} \frac{a_i}{3^i}$ where $a_i \in \{0,2\}\}$.

(6) Show that if K is the middle-thirds Cantor set, then $K + K = [0,2]$, where $K + K = \{z : z = x+y \text{ for some } x, y \in K\}$.

(7) Show that the function ϕ in the proof of Theorem 2.2.2 is onto K minus the endpoints of $F_n, n \geq 1$.

* (8) A real number x is said to be a *Liouville number* if it is irrational and for any integer $n > 0$ there exist integers p and $q > 1$ such that

$$|x - \frac{p}{q}| < \frac{1}{q^n}.$$

Show that the set of Liouville numbers is a null set.

* (9) (The Cantor function) Construct a function $\psi : [0,1] \to [0,1]$ that is continuous, monotone nondecreasing (i.e., if $x \leq y$, then $f(x) \leq f(y)$) in $[0,1]$ and such that ψ is constant on each interval in the complement of K and $\psi(K) = [0,1]$. This is called the Cantor ternary function. (Hint: Represent x in ternary form and $\psi(x)$ in binary form. The value of ψ on the intervals G_n should be constant.)

2.3. Lebesgue Measurable Sets

A set A in $\mathbb{R}$ is said to be **Lebesgue measurable** or **measurable** if for any $\varepsilon > 0$ there is an open set $G = G_\varepsilon$ such that

$$A \subset G \text{ and } \lambda^*(G \setminus A) < \varepsilon.$$

Informally, we see from the definition that measurable sets are those that are "well-approximated" from above by open sets. We first obtain some examples of measurable sets.

Proposition 2.3.1. *Open sets and null sets are measurable.*

Proof. If A is open, then for any $\varepsilon > 0$, G can be taken to be A, so open sets are clearly measurable. Now let N be a null set, i.e., suppose $\lambda^*(N) = 0$. Then for any $\varepsilon > 0$ there exists a sequence of open intervals $\{I_j\}$ whose union covers N and such that $\sum_{j=1}^\infty |I_j| < \varepsilon$. Then $G = \bigcup_{j=1}^\infty I_j$ is open and satisfies

$$\lambda^*(G \setminus N) \leq \lambda^*(G) \leq \sum_{j=1}^\infty |I_j| < \varepsilon.$$

Hence N is measurable. □

Our goal in this section is to show that countable unions, countable intersections and complements of measurable sets are measurable. Together with the fact that open sets and null sets are measurable, this yields a large class of subsets of the line that are measurable; the sets in this class will be defined later as the Lebesgue measurable sets.

Showing that the countable union of measurable sets is measurable is a rather straightforward consequence of the definition, as the following proof of Proposition 2.3.2 shows. The proof for countable intersections is more difficult, and for us it will follow from Proposition 2.3.2 and the fact that complements of measurable sets are measurable. The crucial part for this last fact is to show that closed sets are measurable. In the process, we also obtain a useful characterization of measurable sets (Lemma 2.3.7).

Proposition 2.3.2. *The countable union of measurable sets is measurable.*

Proof. Let $\{A_n\}_{n\geq 1}$ be a sequence of measurable sets and write $A = \bigcup_{n=1}^{\infty} A_n$. Let $\varepsilon > 0$. For each n there exists an open set G_n such that $A_n \subset G_n$ and
$$\lambda^*(G_n \setminus A_n) < \frac{\varepsilon}{2^n}.$$
Let $G = \bigcup_{n=1}^{\infty} G_n$; then G is open and covers A. Since
$$G \setminus A \subset \bigcup_{n=1}^{\infty} (G_n \setminus A_n),$$
then, using countable subadditivity,
$$\lambda^*(G \setminus A) \leq \lambda^*(\bigcup_{n=1}^{\infty} (G_n \setminus A_n)) \leq \sum_{n=1}^{\infty} \lambda^*(G_n \setminus A_n) < \sum_{n=1}^{\infty} \frac{\varepsilon}{2^n} = \varepsilon.$$
Therefore A is measurable. $\square$

The following lemma is an important fact that is true in greater generality (see Theorem 2.4.1), but we need this special case here to prove that closed sets are measurable.

Lemma 2.3.3. *Let $\{E_j\}_{j=1}^{N}$ be a finite collection of disjoint closed bounded sets. Then*
$$\lambda^*(\bigsqcup_{j=1}^{N} E_j) = \sum_{j=1}^{N} \lambda^*(E_j).$$

Proof. By countable subadditivity we only need to show that
$$\lambda^*(\bigsqcup_{j=1}^{N} E_j) \geq \sum_{j=1}^{N} \lambda^*(E_j).$$

2.3. Lebesgue Measurable Sets

First assume $N = 2$; the general case follows immediately by induction. By Lemma B.1.6 there exists a number $\delta > 0$ so that every interval of length less than δ cannot have a nonempty intersection with both E_1 and E_2. As $E_1 \cup E_2$ is bounded, $\lambda^*(E_1 \cup E_2) < \infty$. Let $\varepsilon > 0$. Use Proposition 2.1.1 to find a sequence of intervals $\{I_j\}_{j\geq 1}$, with $|I_j| < \delta$, whose union covers $E_1 \cup E_2$ and such that

$$\sum_{j=1}^\infty |I_j| < \lambda^*(E_1 \cup E_2) + \varepsilon.$$

Let $\Gamma = \{j \geq 1 : I_j \cap E_1 \neq \emptyset\}$. Then

$$E_1 \subset \bigcup_{j \in \Gamma} I_j \text{ and } E_2 \subset \bigcup_{j \in \Gamma^c} I_j.$$

Therefore,

$$\lambda^*(E_1) + \lambda^*(E_2) \leq \sum_{j \in \Gamma} |I_j| + \sum_{j \in \Gamma^c} |I_j| = \sum_{j=1}^\infty |I_j| < \lambda^*(E_1 \cup E_2) + \varepsilon.$$

As this holds for all $\varepsilon > 0$, then $\lambda^*(E_1) + \lambda^*(E_2) \leq \lambda^*(E_1 \cup E_2)$ and this completes the proof. □

We now prove a technical lemma to be used in Lemma 2.3.5 but also of interest in its own right (a generalization of this lemma is offered in Exercise 2.4.2). The new idea in Lemma 2.3.4 is that the intervals are not necessarily disjoint but are allowed to intersect at their endpoints. While the lemma is stated for any bounded intervals, we will only apply it in the case of bounded closed intervals.

Lemma 2.3.4. *Let $\{I_j\}_{j=1}^N$ be a finite collection of bounded intervals that are nonoverlapping. Then*

$$\lambda^*(\bigcup_{j=1}^N I_j) = \sum_{j=1}^N \lambda^*(I_j).$$

Proof. Since the intervals $\{I_j\}$ are nonoverlapping, for each $\varepsilon > 0$ there exists a closed interval $I'_j \subset I_j$ such that $\lambda^*(I'_j) \geq \lambda^*(I_j) - \frac{\varepsilon}{N}$, with $\{I'_j\}$ disjoint. Using Lemma 2.3.3 we obtain that

$$\lambda^*(\bigcup_{j=1}^{N} I_j) \geq \lambda^*(\bigsqcup_{j=1}^{N} I'_j) = \sum_{j=1}^{N} \lambda^*(I'_j)$$

$$\geq \sum_{j=1}^{N} \lambda^*(I_j) - \varepsilon.$$

Letting $\varepsilon \to 0$, $\lambda^*(\bigcup_{j=1}^{N} I_j) \geq \sum_{j=1}^{N} \lambda^*(I_j)$. The reverse inequality follows from subadditivity. □

Lemma 2.3.5 is true in greater generality (see Corollary 2.4.2) but we need it now to show that closed sets are measurable.

Lemma 2.3.5. *If F is a bounded closed set and G is an open set such that $F \subset G$, then*

$$\lambda^*(G \setminus F) = \lambda^*(G) - \lambda^*(F).$$

Proof. Since $G = (G \setminus F) \cup F$ and $\lambda^*(F) < \infty$, by countable subadditivity $\lambda^*(G \setminus F) \geq \lambda^*(G) - \lambda^*(F)$. To prove the other inequality we observe that $G \setminus F$ is open, so there exists a sequence of nonoverlapping, closed bounded intervals $\{I_j\}$ such that $G \setminus F = \bigcup_{j=1}^{\infty} I_j$. Thus, for all $N > 1$,

$$G \supset (\bigcup_{j=1}^{N} I_j) \sqcup F.$$

As both $\bigcup_{j=1}^{N} I_j$ and F are disjoint, bounded and closed, applying Lemma 2.3.3 and then Lemma 2.3.4, we obtain that

$$\lambda^*(G) \geq \lambda^*(\bigcup_{j=1}^{N} I_j) + \lambda^*(F)$$

$$= \sum_{j=1}^{N} \lambda^*(I_j) + \lambda^*(F).$$

Since this holds for all N,

$$\lambda^*(G) \geq \sum_{j=1}^{\infty} \lambda^*(I_j) + \lambda^*(F) \geq \lambda^*(G \setminus F) + \lambda^*(F).$$

□

2.3. Lebesgue Measurable Sets

We are now ready to prove the following proposition. This, together with Lemma 2.3.7, is used later to prove that the complement of a measurable set is measurable.

Proposition 2.3.6. *Closed sets are measurable.*

Proof. Let F be a closed set. First assume that F is bounded, hence of finite outer measure. Thus, for $\varepsilon > 0$ there exist open intervals I_j such that $F \subset \bigcup_{j=1}^{\infty} I_j$ and $\lambda^*(\bigcup_{j=1}^{\infty} I_j) < \lambda^*(F)+\varepsilon$. Let $G = \bigcup_{j=1}^{\infty} I_j$. Then G is open and contains F and by Lemma 2.3.5 $\lambda^*(G \setminus F) = \lambda^*(G) - \lambda^*(F) < \varepsilon$, so F is measurable.

In the general case, write $F_n = F \cap [-n, n]$. Then F_n is closed and bounded and therefore measurable. By Proposition 2.3.2, the union $\bigcup_{n=1}^{\infty} F_n = F$ is measurable. □

The following lemma tells us that a measurable set is "almost" a countable intersection of open sets; its converse is also true and is a consequence of Theorem 2.3.8. A set that is a countable intersection of open sets, such as the set G^* in Lemma 2.3.7, is called a $\mathcal{G}_\delta$ **set**. (In this notation $\mathcal{G}$ stands for open and δ for intersection.)

Lemma 2.3.7. *If a set A is measurable, then there exists a $\mathcal{G}_\delta$ set G^* and a null set N such that*
$$A = G^* \setminus N.$$

Proof. Let A be measurable. For each $\varepsilon_n = \frac{1}{n} > 0$ there exists an open set G_n with $A \subset G_n$ and $\lambda^*(G_n \setminus A) < \varepsilon_n$. Let $G^* = \bigcap_n G_n$. Then $A \subset G^*$, and for all n,
$$\lambda^*(G^* \setminus A) \leq \lambda^*(G_n \setminus A) < \frac{1}{n}.$$
Thus $\lambda^*(G^* \setminus A) = 0$. If $N = G^* \setminus A$, then $A = G^* \setminus N$ and this completes the proof. □

We now state the main result of this section. Parts (1) and (2) have already been shown but are mentioned again for completeness.

Theorem 2.3.8. *The collection of measurable sets satisfies the following properties.*

(1) *The empty set and the set of reals $\mathbb{R}$ are measurable.*

(2) *A countable union of measurable sets is measurable.*

(3) *The complement of a measurable set is measurable.*

(4) *A countable intersection of measurable sets is measurable.*

Proof. Parts (1) and (2) have already been shown in Proposition 2.3.1 and Proposition 2.3.2. For part (3), let A be a measurable set. We know that
$$A = G^* \setminus N,$$
where G^* is a $\mathcal{G}_\delta$ set and N is a null set. Write $G^* = \bigcap_{n=1}^\infty G_n$, where the sets G_n are open. Then
$$A^c = (G^* \cap N^c)^c = (\bigcap_{n=1}^\infty G_n)^c \cup N = \bigcup_{n=1}^\infty G_n^c \cup N.$$
Since the sets G_n^c are closed, and N is measurable, then it follows that A^c is measurable.

Part (4) follows from De Morgan's laws (see Exercise A.6). Indeed, if A_n is a sequence of measurable sets, $(\bigcap_{n=1}^\infty A_n)^c = \bigcup_{n=1}^\infty A_n^c$, which is measurable by part (2). □

We end with another useful characterization of measurable sets.

Lemma 2.3.9. *A set A is measurable if and only if for any $\varepsilon > 0$ there is a closed set F such that $F \subset A$ and $\lambda^*(A \setminus F) < \varepsilon$.*

Proof. Let A be measurable. By Theorem 2.3.8(3) the set A^c is measurable, so for $\varepsilon > 0$ there exists an open set G such that $A^c \subset G$ and $\lambda^*(G \setminus A^c) < \varepsilon$. One can verify that $G^c \subset A$ and $G \setminus A^c = A \setminus G^c$. Then the set $F = G^c$ is closed and $\lambda^*(A \setminus F) < \varepsilon$. For the converse note that if A satisfies the condition of the lemma, then a similar argument shows that A^c is measurable, so A must be measurable. □

Exercises

(1) Show that if A is measurable, then for any t in $\mathbb{R}$, $A + t = \{a + t : a \in A\}$ and $tA = \{ta : a \in A\}$ are measurable. Conclude that then $\lambda(A+t) = \lambda(A)$ and $\lambda(tA) = |t|\lambda(A)$.

(2) Show that if A is a null set, then $A^2 = \{a^2 : a \in A\}$ is also a null set.

2.4. Countable Additivity

(3) Let A be any set. Show that if there is a measurable set B that differs from A by a null set, i.e., $\lambda^*(A \triangle B) = 0$, then A is measurable.

(4) Show the converse of Lemma 2.3.7, i.e., show that if $A = G^* \setminus N$ where G^* is a $\mathcal{G}_\delta$ set and N is a null set, then A is measurable.

(5) Show that a set A is measurable if and only if there exist a set F^* and a set N such that F^* is a countable union of closed sets, N is a null set and $A = F^* \cup N$. A set that is a countable union of closed sets is called an $\mathcal{F}_\sigma$ **set.** (In this notation $\mathcal{F}$ stands for closed and σ for union.)

(6) Show that every closed set is a $\mathcal{G}_\delta$ and every open set is an $\mathcal{F}_\sigma$.

(7) Show that A is measurable if and only if for any $\varepsilon > 0$ there is a closed set F and an open set G such that $F \subset A \subset G$ and $\lambda^*(G \setminus F) < \varepsilon$.

(8) Let A be a bounded set. Show that A is measurable if and only if for any $\varepsilon > 0$ there is a closed set F such that $F \subset A$ and $\lambda^*(F) > \lambda^*(A) - \varepsilon$. What is the difference between this characterization and Lemma 2.3.9?

* (9) (*Carathéodory's Criterion*) Show that a set A is a measurable set if and only if for any set B it is the case that $\lambda^*(B) = \lambda^*(B \cap A) + \lambda^*(B \cap A^c)$. (This is studied in the context of arbitrary measure spaces in Section 2.8.)

2.4. Countable Additivity

When restricted to the Lebesgue measurable sets $\mathfrak{L}$, Lebesgue outer measure λ^* is denoted by λ and called the **Lebesgue measure** on $\mathbb{R}$; so $\lambda(A) = \lambda^*(A)$ for all measurable sets A.

We next show one of the most important properties of Lebesgue measure. This property is called countable additivity and is what characterizes a *measure* as defined in the next section.

Theorem 2.4.1 (Countable Additivity). *If $\{A_n\}_{n=1}^{\infty}$ is a sequence of disjoint measurable sets, then*

$$\lambda(\bigsqcup_{n=1}^{\infty} A_n) = \sum_{n=1}^{\infty} \lambda(A_n).$$

Proof. Let

$$A = \bigsqcup_{n=1}^{\infty} A_n.$$

By countable subadditivity, it suffices to show the following inequality:

$$\sum_{n=1}^{\infty} \lambda(A_n) \leq \lambda(A).$$

We do this first for the case when A is bounded (in fact, we only use that each A_n is bounded). In this case, by Lemma 2.3.9, for each $\varepsilon > 0$ and $n \geq 1$ there exists a closed set $F_n \subset A_n$ such that $\lambda(A_n \setminus F_n) < \frac{\varepsilon}{2^n}$. As $A_n = (A_n \setminus F_n) \cup F_n$ and $\lambda(A_n) < \infty$, we have

$$\lambda(A_n) < \lambda(F_n) + \frac{\varepsilon}{2^n}.$$

So for every integer $N > 1$,

$$\sum_{n=1}^{N} \lambda(A_n) < \sum_{n=1}^{N} \lambda(F_n) + \varepsilon.$$

Since the sets F_n are disjoint, closed and bounded, using Lemma 2.3.3,

$$\lambda(\bigsqcup_{n=1}^{N} F_n) = \sum_{n=1}^{N} \lambda(F_n).$$

Therefore,

$$\sum_{n=1}^{N} \lambda(A_n) < \lambda(\bigsqcup_{n=1}^{N} F_n) + \varepsilon \leq \lambda(A) + \varepsilon.$$

Taking limits as $N \to \infty$ we obtain

$$\sum_{n=1}^{\infty} \lambda(A_n) \leq \lambda(A) + \varepsilon.$$

Letting $\varepsilon \to 0$ completes the proof of this part.

2.4. Countable Additivity

Finally, if A is not bounded, then for any integer i and any $n \geq 1$ write
$$B_{n,i} = A_n \cap [i, i+1).$$
It follows that for each i, $[i, i+1) \cap A = \bigsqcup_{n=1}^{\infty} B_{n,i}$, a disjoint union. Since $[i, i+1) \cap A$ is bounded, the first part yields
$$\lambda([i, i+1) \cap A) = \sum_{n=1}^{\infty} \lambda(B_{n,i}).$$
Now for any $N > 0$,
$$\lambda(A) \geq \lambda\bigl(\bigsqcup_{i=-N}^{N} [i, i+1) \cap A\bigr) = \sum_{i=-N}^{N} \lambda([i, i+1) \cap A).$$
Therefore,
$$\lambda(A) \geq \sum_{i=-\infty}^{\infty} \lambda([i, i+1) \cap A) = \sum_{i=-\infty}^{\infty} \sum_{n \geq 1} \lambda(B_{i,n})$$
$$= \sum_{n=1}^{\infty} \sum_{i=-\infty}^{\infty} \lambda(B_{i,n}) = \sum_{n=1}^{\infty} \lambda(A_n),$$
which completes the proof. □

The following is a useful corollary whose proof is left to the reader.

Corollary 2.4.2. *Let A and B be measurable sets such that $B \subset A$ and $\lambda(B) < \infty$. Then $\lambda(A \setminus B) = \lambda(A) - \lambda(B)$.*

Exercises

(1) Prove Corollary 2.4.2. Show that if A and B are measurable sets with $\lambda(B) < \infty$, then $\lambda(A \setminus B) \geq \lambda(A) - \lambda(B)$.

(2) Define a sequence of measurable sets $\{A_n\}_{n \geq 1}$ to be **almost disjoint** if $\lambda(A_n \cap A_m) = 0$ for all $n \neq m$. Show that if $\{A_n\}_{n \geq 1}$ are almost disjoint, then $\lambda(\bigcup_{n \geq 1} A_n) = \sum_{n \geq 1} \lambda(A_n)$.

(3) Let A be a set of finite outer measure. Show that the set A is measurable if and only if for any $\varepsilon > 0$ there is a set H such that H is a finite union of bounded intervals and $\lambda^*(A \triangle H) < \varepsilon$.

(4) Show that any collection of disjoint sets of positive measure is countable. (Note that, as we saw in Section 2.2, positive measure sets need not contain intervals.)

(5) Modify the construction of the Cantor middle-thirds set in the following way. At the first stage remove a central interval of length $\frac{1}{6}$ and at the nth stage, instead of removing the open middle-thirds, remove slightly smaller intervals so that in the end the total measure of the removed intervals is $\frac{1}{2}$. Show that in this way you obtain a Cantor set that is not a null set. What is the measure of this set?

(6) Modify the construction of the previous exercise to obtain a set of measure α for any $0 \leq \alpha < 1$ and show that this set is bounded, perfect and totally disconnected.

(7) (Measure-theoretic union) Let $\{A_\alpha\}_{\alpha \in \Gamma}$ be an arbitrary collection of measurable sets. Show that there exists a measurable set A (called a *measure-theoretic union*) such that $A \subset \bigcup A_\alpha$ and $\lambda(A_\alpha \setminus A) = 0$ for all $\alpha \in \Gamma$. (Hint: First assume that all sets are in $[0, 1]$ and consider the collection of all countable unions of elements from $\{A_\alpha\}_{\alpha \in \Gamma}$.)

2.5. Sigma-Algebras and Measure Spaces

The unit interval with Lebesgue measure is the prototype of a finite measure space. It is the most important measure space that we study. The collection of Lebesgue measurable subsets of the unit interval is one of the important examples of a σ-algebra.

Theorem 2.3.8 states that the collection of Lebesgue measurable sets is a nonempty collection of subsets of $\mathbb{R}$ that is closed under countable unions, countable intersections and complements. A nonempty collection of sets with these properties is called a *σ-algebra*. This concept plays a crucial role in the development of the general theory of measure. While we give the general definition, our emphasis will be on three kinds of σ-algebras: the Lebesgue measurable sets in $\mathbb{R}$, the Borel sets in $\mathbb{R}$, and the collection of all subsets of a finite or countable set X. As we shall see, each of these σ-algebras induces a

2.5. Sigma-Algebras and Measure Spaces

σ-algebra on the subsets of a Lebesgue set, a Borel set or a subset of X.

Let X be a nonempty set (usually a measurable subset of $\mathbb{R}$ or $\mathbb{R}^d$). A **σ-algebra** on X is a collection $\mathcal{S}$ of subsets of X such that

(1) $\mathcal{S}$ is nonempty;

(2) $\mathcal{S}$ is closed under complements, i.e., whenever $A \in \mathcal{S}$, then $A^c \in \mathcal{S}$ (here $A^c = X \setminus A$);

(3) $\mathcal{S}$ is closed under countable unions, i.e., whenever $A_n \in \mathcal{S}, n \geq 1$, then $\bigcup_{n=1}^{\infty} A_n \in \mathcal{S}$.

Since $\bigcap_{n=1}^{\infty} A_n = (\bigcup_{n=1}^{\infty} A_n^c)^c$, a σ-algebra is closed under countable intersections. As $\mathcal{S}$ must contain at least one element, say A, and $X = A \cup A^c$, it follows that X, and hence $\emptyset$, are always in $\mathcal{S}$. We have already seen that the collection of Lebesgue measurable subsets of $\mathbb{R}$ is a σ-algebra. There are two σ-algebras that exist on any set X. The first is $\{\emptyset, X\}$, the smallest σ-algebra of subsets of X, called the **trivial σ-algebra**. The other one is the collection of all subsets of X or the **power set** of X, denoted by $\mathcal{P}(X)$, the largest σ-algebra (i.e., contains any other σ-algebra) of subsets of X, called the **improper σ-algebra**. The improper σ-algebra on a set X will be of interest when X is a finite or countable set.

Question. Show that if $\mathcal{S}$ is a collection of subsets of a set X that contains X and is closed under set-differences (i.e., if $A, B \in \mathcal{S}$, then $A \setminus B \in \mathcal{S}$) and countable unions, then $\mathcal{S}$ is a σ-algebra.

Example. Let $X = \{a, b, c, d\}$. Then the collection $\mathcal{S}$ defined by

$$\mathcal{S} = \{\emptyset, X, \{a\}, \{b, c, d\}, \{a, b\}, \{c, d\}, \{a, c, d\}, \{b\}\}$$

is a σ-algebra on X. Also, if we set $Y = \{a, c, d\}$ and let $\mathcal{S}(Y)$ denote the collection $\{A \cap Y : A \in \mathcal{S}\}$, then

$$\mathcal{S}(Y) = \{\emptyset, Y, \{a\}, \{c, d\}\}$$

is a σ-algebra on Y. Also note that $\mathcal{S}(Y) = \{A : A \subset Y \text{ and } A \in \mathcal{S}\}$. Now, if we let $Z = \{a, b, c\}$, then Z is not in $\mathcal{S}$ and

$$\mathcal{S}(Z) = \{\emptyset, Z, \{a\}, \{b, c\}, \{a, b\}, \{c\}, \{a, c\}, \{b\}\}$$

is also a σ-algebra on Z. But note that in this case the collection $\{A : A \subset Z \text{ and } A \in \mathcal{S}\} = \{\emptyset, \{a\}, \{a,b\}, \{b\}\}$ is not a σ-algebra on Z.

The proof of the following proposition is left to the reader.

Proposition 2.5.1. *Let X be a nonempty set and let $\mathcal{S}$ be a σ-algebra on X.*

(1) *If $Y \subset X$, then the collection of* **sets restricted to** Y *defined by $\mathcal{S}(Y) = \{A \cap Y : A \in \mathcal{S}\}$ is a σ-algebra on Y. ($\mathcal{S}(Y)$ is also denoted by $\mathcal{S} \cap Y$.)*

(2) *If $Y \in \mathcal{S}$, then*
$$\mathcal{S}(Y) = \{A : A \subset Y \text{ and } A \in \mathcal{S}\}.$$

The collection of all Lebesgue measurable sets in $\mathbb{R}$ is denoted by $\mathfrak{L}$. If X is a Lebesgue measurable subset of $\mathbb{R}$, the set of Lebesgue measurable sets contained in X is denoted by $\mathfrak{L}(X)$. By Proposition 2.5.1 $\mathfrak{L}(X)$ is a σ-algebra of subsets of X.

It is important to note that there exist subsets of the reals that are not Lebesgue measurable; a construction of a nonmeasurable set is given in Section 3.2. In fact, it can further be shown that every set of positive Lebesgue outer measure contains a non-Lebesgue measurable subset (see Exercise 3.11.1 or Oxtoby [**56**, Ch. 5]).

Lebesgue measure is defined on the σ-algebra of measurable sets. As we shall see, sometimes we may find it useful to consider other "measures" such as a multiple of Lebesgue measure. The important property that Lebesgue measure enjoys is that it is countably additive. To make this precise we introduce the following definition. Let X be a nonempty set and $\mathcal{S}$ a σ-algebra on X. A **measure on** $\mathcal{S}$ is a function μ defined on $\mathcal{S}$ and with values in $[0, \infty]$ that satisfies the following two properties:

(1) $\mu(\emptyset) = 0$;

(2) μ is countably additive: for any collection of disjoint sets $\{A_n\}_{n \geq 1}$ in $\mathcal{S}$,
$$\mu(\bigsqcup_{n=1}^{\infty} A_n) = \sum_{n=1}^{\infty} \mu(A_n).$$

2.5. Sigma-Algebras and Measure Spaces

Evidently, if X is a Lebesgue measurable set and $\mathcal{S} = \mathfrak{L}(X)$, then λ is a measure on $\mathcal{S}$. But, for example, 2λ is also a measure on $\mathcal{S}$, and sometimes we may want to consider a "normalized" measure such as $\frac{1}{2}\lambda$ on $\mathfrak{L}([-1, 1])$.

While we mainly will be concerned with Lebesgue spaces, there are many situations where the main property of the space that is used is the countable additivity of the measure. So we introduce the notion of a *measure space*.

A **measure space** is defined to be a triple $(X, \mathcal{S}, \mu)$ where X is a nonempty set, $\mathcal{S}$ is a σ-algebra in X and μ is a measure on $\mathcal{S}$. The elements of $\mathcal{S}$ are sets that are said to be measurable with respect to $\mathcal{S}$ or $\mathcal{S}$-measurable. A **probability space** is a measure space $(X, \mathcal{S}, \mu)$ such that $\mu(X) = 1$.

A measure space $(X, \mathcal{S}, \mu)$ is a **finite measure space** if $\mu(X) < \infty$ and it is **σ-finite** if there is a sequence of measurable sets A_n of finite measure such that $X = \bigcup_{n=1}^{\infty} A_n$. A σ-finite measure space may be of finite measure. Whenever we consider a measure space that has infinite measure we shall always assume it is σ-finite. Evidently, the real line with Lebesgue sets and Lebesgue measure is a σ-finite measure space, and the reader is asked to verify that all canonical Lebesgue measure spaces are σ-finite.

Another property that is enjoyed by Lebesgue measure is that of being *complete*. A measure space $(X, \mathcal{S}, \mu)$ is called **complete** if whenever $A \in \mathcal{S}$ and $\mu(A) = 0$, then for every $B \subset A$ we have $B \in \mathcal{S}$ (so $\mu(B) = 0$). In other words, in a complete measure space every null set (a set contained in a set of measure 0) is measurable.

We now define an important class of measure spaces. A **canonical nonatomic Lebesgue measure space** is a measure space $(X_0, \mathfrak{L}(X_0), \lambda)$ where X_0 is a bounded (of positive length) or unbounded interval in $\mathbb{R}$ and λ is Lebesgue measure on $\mathfrak{L}(\mathbb{R})$.

The next important class of examples are the **canonical atomic Lebesgue spaces** that we now define. Let Z be any nonempty subset of $\mathbb{Z}$ and let the σ-algebra $\mathcal{S}$ be $\mathcal{P}(Z)$; if Z is a finite set let $\#(Z)$ denote its number of elements. The main examples that we consider are when Z is $Z_n = \{0, \ldots, n-1\}$, or $\mathbb{Z}$, or $\mathbb{N}_0$. For each $k \in Z$ define

$$\nu(\{k\}) = \begin{cases} 1/\#(Z), & \text{if } Z \text{ is finite}; \\ 1, & \text{if } Z \text{ is infinite}, \end{cases}$$

$$\nu(\emptyset) = 0,$$

and extend ν to $\mathcal{S}$ by $\nu(A) = \sum_{k \in A} \nu(\{k\})$. The reader should verify that ν is countably additive on $\mathcal{S}$. The sets $\{k\}$ are called the **atoms** of the space, and the elements of $\mathcal{P}(Z)$ are the measurable sets of the space; every measurable set of the space is a countable union of atoms. The measure ν on Z_n can be thought of as modeling the tosses of a fair n-sided die. When $Z = \mathbb{Z}$ we call ν a **counting measure**.

A **canonical Lebesgue measure space** is defined to be a triple $(X, \mathcal{S}(X), \mu)$ such that

$$X = X_0 \sqcup Z \text{ or } X = X_0 \text{ or } X = Z,$$

where $(X_0, \mathfrak{L}(X_0), \lambda)$ is a canonical nonatomic Lebesgue space, and $(Z, \mathcal{P}(Z), \nu)$ is a canonical atomic Lebesgue space. The σ-algebra on X is given by $\mathcal{S}(X) = \{A \sqcup B : A \in \mathfrak{L}(X_0), B \in \mathcal{P}(Z)\}$, and μ is defined by

$$\mu(A) = \lambda(A \cap X_0) + \nu(A \cap Z) \text{ for } A \in \mathcal{S}(X).$$

Question. Show that $\mathcal{S}(X)$ as defined above is a σ-algebra on X. In Exercise 2 the reader is asked to show that μ is a measure on $\mathcal{S}(X)$.

A *Lebesgue space* will be defined later to be any measure space *isomorphic* to a canonical Lebesgue measure space.

The following exercise and proposition are two examples of statements that only use the countable additivity property of the measure μ, and thus hold in the more general setting of measure spaces.

Question. Let $(X, \mathcal{S}, \mu)$ be a measure space. Show that if A, B are measurable sets with $A \subset B$, then $\mu(A) \leq \mu(B)$, and if $\mu(A) < \infty$, then $\mu(B \setminus A) = \mu(B) - \mu(A)$.

2.5. Sigma-Algebras and Measure Spaces

Proposition 2.5.2. *Let $(X, \mathcal{S}, \mu)$ be a measure space.*

(1) *If $\{A_n\}_{n \geq 1}$ is a sequence of measurable sets in X that is increasing, i.e.,*
$$A_n \subset A_{n+1} \quad \text{for all } n \geq 1,$$
then
$$\mu(\bigcup_{n=1}^{\infty} A_n) = \lim_{n \to N} \mu(A_n).$$

(2) *If $\{B_n\}_{n \geq 1}$ is a sequence of measurable sets in X that is decreasing, i.e.,*
$$B_n \supset B_{n+1} \quad \text{for all } n \geq 1,$$
and $\mu(B_K) < \infty$ for some $K > 0$, then
$$\mu(\bigcap_{n=1}^{\infty} B_n) = \lim_{n \to \infty} \mu(B_n).$$

Proof. To prove part (1), let $A = \bigcup_{n=1}^{\infty} A_n$. If $\mu(A_n) = \infty$ for some $n > 0$, then $\mu(A) = \infty$ and $\mu(A_m) = \infty$ for all $m \geq n$, and we are done; so assume that $\mu(A_n) < \infty$ for all $n > 0$. Write $A_0 = \emptyset$ and observe that
$$A = \bigsqcup_{n=0}^{\infty} (A_{n+1} \setminus A_n).$$
The sets in the above union are disjoint as the A_n are increasing. Thus, by countable additivity,
$$\mu(A) = \sum_{n=0}^{\infty} \mu(A_{n+1} \setminus A_n) = \lim_{N \to \infty} \sum_{n=0}^{N} \mu(A_{n+1} \setminus A_n)$$
$$= \lim_{N \to \infty} \sum_{n=0}^{N} \mu(A_{n+1}) - \mu(A_n) = \lim_{N \to \infty} \mu(A_{N+1}),$$
which completes the proof of part a).

To prove part (2) note that since $\bigcap_{n=1}^{\infty} B_n = \bigcap_{n=K}^{\infty} B_n$, after renaming the sets if necessary we may assume that $\mu(B_n) < \infty$ for all $n \geq 1$ (so $K = 1$). Write
$$C = \bigcap_{n=1}^{\infty} B_n.$$

Now note that

$$(2.4) \qquad B_1 = C \sqcup \bigsqcup_{n=1}^{\infty} (B_n \setminus B_{n+1}).$$

Since the sets in (2.4) are disjoint, by countable additivity,

$$\mu(B_1) = \mu(C) + \sum_{n=1}^{\infty} \mu(B_n \setminus B_{n+1}).$$

Then we observe that

$$\sum_{n=1}^{\infty} \mu(B_n \setminus B_{n+1}) = \lim_{N \to \infty} \sum_{n=1}^{N-1} (\mu(B_n) - \mu(B_{n+1}))$$
$$= \mu(B_1) - \lim_{N \to \infty} \mu(B_N).$$

Therefore,

$$\mu(B_1) - \mu(C) = \mu(B_1) - \lim_{N \to \infty} \mu(B_N),$$

so

$$\mu(C) = \lim_{N \to \infty} \mu(B_N).$$

This concludes the proof. □

Exercises

(1) Prove Proposition 2.5.1.

(2) Let μ be as in the definition of a canonical Lebesgue space. Show that μ is a measure.

(3) Show that Proposition 2.5.2, part b), does not hold without the assumption that $\mu(B_K) < \infty$ for some $K \geq 1$.

(4) Let A and B be measurable sets such that $\mu(A) < \infty$. Let $\varepsilon > 0$. Show that $\mu(A \setminus B) < \varepsilon$ if and only if $\mu(A \cap B) > \mu(A) - \varepsilon$.

(5) Let A and B be measurable sets such that $\mu(A) < \infty$. Let $\varepsilon > 0$. Show that if $\mu(A \triangle B) < \varepsilon$, then $\mu(A) - \varepsilon < \mu(B) < \mu(A) + \varepsilon$.

(6) (Triangle Inequality) Let A, B, C be measurable sets. Show that

$$\mu(A \triangle B) \leq \mu(A \triangle C) + \mu(C \triangle B).$$

2.5. Sigma-Algebras and Measure Spaces

(7) Show that if μ is a measure on a σ-algebra $\mathcal{S}$ of some set X, then it must be countably subadditive, i.e., for any sets $A_n \in \mathcal{S}$, $\mu(\bigcup_{n=1}^{\infty} A_n) \leq \sum_{n=1}^{\infty} \mu(A_n)$.

(8) Show that a canonical Lebesgue measure space is a σ-finite and complete measure space.

(9) Let $X = \mathbb{R}$ and let $\mathcal{S}$ be the collections of all subsets of X. For $A \subset X$ define $\mu(A)$ to be the number of elements in A if A is finite, and ∞ otherwise. Show that $(X, \mathcal{S}, \mu)$ is a measure space that is not σ-finite.

(10) Let X be a Lebesgue measurable set. Given any two Lebesgue measurable sets A and B in X, define the relation $A \sim B$ when $\lambda(A \triangle B) = 0$. Show that $\sim$ is an equivalent relation on the elements of $\mathfrak{L}(X)$.

(11) Let $\{A_n\}, n \geq 1$, be a sequence of subsets of a set X. Define

$$\liminf_{n \to \infty} A_n = \bigcup_{m=1}^{\infty} \bigcap_{n=m}^{\infty} A_n,$$

$$\limsup_{n \to \infty} A_n = \bigcap_{m=1}^{\infty} \bigcup_{n=m}^{\infty} A_n.$$

Show that $\liminf_{n \to \infty} A_n$ consists of the sets of points $x \in X$ that are in A_n for all large n (we say that they are eventually in A_n), and that $\limsup_{n \to \infty} A_n$ consists of the sets of points $x \in X$ that are in infinitely many A_n.

(12) (Borel–Cantelli) Let $(X, \mathcal{S}, \mu)$ be a measure space and let $\{A_n\}$ be a sequence of measurable sets. Show that if $\sum_{n=1}^{\infty} \mu(A_n) < \infty$, then $\mu(\limsup A_n) = 0$.

(13) Let $[A]$ denote the equivalence class of the measurable set A under the equivalence relation of Exercise 10. Let $\mathfrak{L}(X)/\sim$ denote the set of equivalence classes, so $\mathfrak{L}(X)/\sim = \{[A] : A \in \mathfrak{L}(X)\}$. For any two equivalence classes $[A]$ and $[B]$ define $d([A], [B]) = \lambda(A \triangle B)$. Show that d is a metric on $\mathfrak{L}(X)/\sim$.

The following exercises develop the notion of *completion* of a measure space.

(14) Let $(X, \mathcal{S}, \mu)$ be a measure space. Define the **completion** of $\mathcal{S}$ with respect to μ to be the collection of sets $\mathcal{S}_\mu$ consisting of all sets $E \subset X$ such that there exist $A, B \in \mathcal{S}$ with

$$A \subset E \subset B \text{ and } \mu(B \setminus A) = 0.$$

Show that $\mathcal{S}_\mu$ is a σ-algebra containing $\mathcal{S}$.

(15) Let $(X, \mathcal{S}, \mu)$ be a measure space and let $\mathcal{S}_\mu$ be as in Exercise 14. Define $\bar{\mu}$ on elements of $\mathcal{S}_\mu$ by $\bar{\mu}(E) = \mu(A)$ for any $A \in \mathcal{S}$ such that there is a $B \in \mathcal{S}$ with $A \subset E \subset B$ and $\mu(B \setminus A) = 0$. Show that the value of $\bar{\mu}$ on E is independent of A and B and therefore is a well-defined set function. Furthermore, show that $\bar{\mu}$ is a complete measure on $\mathcal{S}_\mu$. We say that $(X, \mathcal{S}_\mu, \bar{\mu})$ is the (measure) completion of $(X, \mathcal{S}, \mu)$.

2.6. The Borel Sigma-Algebra

This section introduces the notion of a collection of sets generating a σ-algebra and defines the σ-algebra of Borel sets, a σ-algebra properly contained in the σ-algebra of Lebesgue sets.

Given any two σ-algebras $\mathcal{S}_1$ and $\mathcal{S}_2$ on a nonempty set X, their intersection $\mathcal{S}_1 \cap \mathcal{S}_2$, defined by

$$\mathcal{S}_1 \cap \mathcal{S}_2 = \{A : A \in \mathcal{S}_1 \text{ and } A \in \mathcal{S}_2\},$$

is also a σ-algebra. This follows from the simple fact that any collection of sets that belongs to the intersection also belongs to each σ-algebra, so the collection is closed under complements and countable unions; also the intersection contains at least one element, namely X. By a similar reasoning, only with minor changes to take care of the more complicated notation, the intersection of any (countable or not) collection of σ-algebras on a nonempty set X is a σ-algebra. Given any collection of subsets $\mathcal{C}$ of a nonempty set X we define the σ-**algebra generated** by $\mathcal{C}$ to be the intersection of all the σ-algebras containing $\mathcal{C}$; this σ-algebra is denoted by $\sigma(\mathcal{C})$. Note that there is always at least one σ-algebra containing $\mathcal{C}$, namely $\mathcal{P}(X)$, the power set of X. It follows that $\sigma(\mathcal{C})$ is defined for any collection $\mathcal{C}$ and it is a σ-algebra; it is characterized by the fact that if $\mathcal{A}$ is any σ-algebra

2.6. Borel Sigma-Algebra

containing $\mathcal{C}$, then $\sigma(\mathcal{C}) \subset \mathcal{A}$. The reader should verify that we have proved the following lemma.

Lemma 2.6.1. *Let X be a nonempty set and let $\mathcal{C}$ be a collection of subsets of X. Then the σ-algebra generated by $\mathcal{C}$ and denoted $\sigma(\mathcal{C})$ is the unique σ-algebra containing $\mathcal{C}$ and such that for any σ-algebra $\mathcal{A}$ containing $\mathcal{C}$ it is the case that $\sigma(\mathcal{C}) \subset \mathcal{A}$.*

We are ready to define another of the important σ-algebras in analysis. The σ-algebra of **Borel sets** $\mathcal{B}$ in $\mathbb{R}$ is defined to be the σ-algebra generated by the open sets. Recall that $\mathcal{G}$ stands for the collection of open sets, so

$$\mathcal{B} = \sigma(\mathcal{G}).$$

By definition, $\mathcal{B}$ contains the open sets and the closed sets, and one might be tempted to think of the Borel sets as being obtained from the open sets by a countable number of unions, intersections and complements. However, the description of the Borel sets is subtler and the definition we have given of $\mathcal{B}$ is nonconstructive; to specify all of its members one needs to make use of transfinite induction.

To give an idea of the complexity of the Borel sets we introduce some notation. If $\mathcal{A}$ is a collection of sets, we let $\mathcal{A}_\delta$ denote the collection of countable intersections of sets from $\mathcal{A}$, and the collection of all countable unions of sets from $\mathcal{A}$ is denoted by $\mathcal{A}_\sigma$. We let $\mathcal{A}_{\delta\sigma}$ denote $(\mathcal{A}_\delta)_\sigma$, etc. Let $\mathcal{F}$ denote the closed sets (this comes from the French, *fermé*, for closed). Then we have the following proper inclusions for classes of Borel sets:

$$\mathcal{G} \subsetneq \mathcal{G}_\delta \subsetneq \mathcal{G}_{\delta\sigma} \subsetneq \mathcal{G}_{\delta\sigma\delta} \subsetneq \ldots,$$
$$\mathcal{F} \subsetneq \mathcal{F}_\sigma \subsetneq \mathcal{F}_{\sigma\delta} \subsetneq \mathcal{F}_{\sigma\delta\sigma} \subsetneq \ldots.$$

It can be shown that there are Borel sets that are not in the union of the classes above, and to obtain all the Borel sets we need to extend the classes above for all countable ordinals using transfinite induction. We do not do this here, as it is beyond the scope of this book.

All Borel sets are Lebesgue measurable, since the collection of Lebesgue measurable sets is a σ-algebra containing the open sets, and since the Borel σ-algebra is the smallest σ-algebra containing the open sets it must be contained in the Lebesgue σ-algebra. Thus we have seen that $\mathcal{B} \subset \mathcal{L}$. Given a Borel set $X \subset \mathbb{R}$ we will use $\mathcal{B}(X)$ to denote the σ-algebra of Borel sets contained in X. By a similar reasoning, it follows that $\mathcal{B}(X) \subset \mathcal{L}(X)$

Furthermore, it can be shown that there are Lebesgue measurable sets that are not Borel sets. We can give the idea of this argument; it is based on showing that the cardinality of the collection of Lebesgue measurable sets is greater than the cardinality of the collection of Borel sets. Let $\mathbf{c}$ denote the cardinality of the continuum $\mathbb{R}$. The set of all subsets of $\mathbb{R}$ has cardinality $2^\mathbf{c}$, and is greater than $\mathbf{c}$ (by the Schroeder-Berstein theorem of set theory). Therefore there are at most $2^\mathbf{c}$ Lebesgue measurable sets. But there exists an uncountable set K of measure zero, namely the Cantor set K defined in Section 2.2. All subsets of K must be of measure zero. Hence they are measurable. Since there are $2^\mathbf{c}$ subsets of K it follows that there are at least $2^\mathbf{c}$ Lebesgue measurable sets. Therefore there are $2^\mathbf{c}$ Lebesgue measurable sets. Using transfinite induction it can be shown that the cardinality of the collection of Borel sets is $\mathbf{c}$. (We just give an argument showing that the cardinality of the collection of open sets is $\mathbf{c}$. Note that each open set can be written as a countable disjoint union of intervals with rational endpoints, which can be identified with the collection of all subsets of $\mathbb{Q}$, which has cardinality $\mathbf{c}$.) Thus, there are Lebesgue sets that are not Borel. So we have that $\mathcal{B}(\mathbb{R}) \subsetneq \mathcal{L}(\mathbb{R})$. It also follows that the Borel σ-algebra does not contain all sets of measure zero.

As we have mentioned, there is no constructive way to describe the σ-algebra generated by a set. However, there is a useful approach to the generated σ-algebra in terms of monotone classes. A **monotone class** on a nonempty set X is a nonempty collection $\mathcal{M}$ of subsets of X that is closed under countable increasing unions and countable decreasing intersections. As the power set of X is a monotone class, we can define the **monotone class generated by** a collection $\mathcal{C}$ as the intersection of all monotone classes containing $\mathcal{C}$. Instead of considering the monotone class generated by an arbitrary collection, we shall put some additional structure on the generating collection. Define an **algebra** on a nonempty set X to be a nonempty collection

2.6. Borel Sigma-Algebra

of subsets of X that is closed under complements and finite unions. It follows that it must be closed under finite intersections. Of course, any σ-algebra is an algebra, but the collection of all finite unions of intervals in $[0,1]$ is an algebra that is not a σ-algebra. We start with the following elementary lemma.

Lemma 2.6.2. *Let X be a nonempty set. If a monotone class $\mathcal{M}$ is an algebra, then it is a σ-algebra.*

Proof. It suffices to show that $\mathcal{M}$ is closed under countable unions. Let $\{A_n\}$ be a sequence of sets in $\mathcal{M}$. Define $A'_n = \bigcup_{i=1}^n A_i$. Then the sequence $\{A'_n\}$ is monotone increasing, so its union is in $\mathcal{M}$, but as it has the same union as the sequence $\{A_n\}$, it follows that $\bigcup_{n=1}^\infty A_n$ is in $\mathcal{M}$. □

We are ready to prove the theorem that characterizes the generated σ-algebra in terms of monotone classes. While this theorem is useful, its proof is somewhat mysterious. An extension of the theorem where "algebra" is replaced by "ring" appears in Exercise 2.7.12.

Theorem 2.6.3 (Monotone Class Theorem). *Let X be a nonempty set and let $\mathcal{A}$ be an algebra of subsets of X. Then, the σ-algebra generated by $\mathcal{A}$, $\sigma(\mathcal{A})$, is equal to the monotone class generated by $\mathcal{A}$, denoted by $\mathcal{M}$.*

Proof. We show that $\sigma(\mathcal{A})$ is contained in $\mathcal{M}$. The other direction is simpler and left as an exercise. We observe that as $\mathcal{M}$ contains $\mathcal{A}$, by Lemma 2.6.2, it suffices to show that $\mathcal{M}$ is an algebra. For each $E \subset X$ define

$$\mathcal{M}(E) = \{A \subset X : A \cup E, A \setminus E, E \setminus A \in \mathcal{M}\}.$$

As $\mathcal{M}$ is a monotone class, it readily follows that when the collection $\mathcal{M}(E)$ is nonempty, it is a monotone class. Also, if one takes E to be in the algebra $\mathcal{A}$, then $\mathcal{A} \subset \mathcal{M}(E)$. So the monotone class $\mathcal{M}$ must be contained in $\mathcal{M}(E)$. Thus, for any A in $\mathcal{M}$ and any E in $\mathcal{A}$, $A \in \mathcal{M}(E)$, which is equivalent to $E \in \mathcal{M}(A)$. So for any A in $\mathcal{M}$, $\mathcal{M} \subset \mathcal{M}(A)$. This, with the fact that X is in $\mathcal{M}$, means that $\mathcal{M}$ is an algebra and completes the proof. □

Exercises

(1) Show that the intersection of any collection of σ-algebras is a σ-algebra.

(2) Give an example of a $\mathcal{G}_\delta$ set that is neither open nor closed.

(3) Show that every null set is contained in a Borel null set.

(4) Let $\mathcal{A}$ be a collection of subsets of a set X that contains X and is closed under the operation of set difference (i.e., if $A, B \in \mathcal{A}$, then $A \setminus B \in \mathcal{A}$) and countable disjoint unions. Show that $\mathcal{A}$ is a σ-algebra. (All that is needed is to show that $\mathcal{A}$ is closed under countable unions.)

(5) Show that the σ-algebra generated by the collection of all closed intervals with rational endpoints is equal to the Borel σ-algebra.

(6) Show that the Borel σ-algebra is generated by $\mathcal{F}$, the collection of closed sets in $\mathbb{R}$.

(7) Show that $(\mathbb{R}, \mathcal{L}, \lambda)$ is the completion (in the sense of Exercise 2.5.15) of $(\mathbb{R}, \mathcal{B}, \lambda)$.

(8) Show that the set consisting of all open subsets of $\mathbb{R}$ can be put in a one-to-one correspondence with $\mathbb{R}$.

(9) Let X be a Borel subset of $\mathbb{R}$ and let $\mathcal{B}(X)$ denote the σ-algebra of Borel sets contained in X. Show that $(X, \mathcal{B}(X), \lambda)$ is a measure space that is σ-finite but not complete. (You may use that there exists a one-to-one correspondence between the collection of Borel sets and $\mathbb{R}$.)

(10) Show that in the proof of Theorem 2.6.3, $\mathcal{M}$ is contained in $\sigma(\mathcal{A})$.

2.7. Approximation with Semi-rings

Intervals play an important role when approximating Lebesgue measurable sets. In the more general setting, semi-rings replace the collection of intervals. In this section we study notions of approximation in more detail and study approximating collections more general than the collection of all intervals.

2.7. Approximation with Semi-rings

When proving properties of measurable sets, often these properties are first shown for intervals or for finite unions of intervals, and then an approximation argument is used to extend the property to all measurable sets. In many cases, however, it will be convenient or necessary to consider a collection different from the collection of all intervals. For example, one may consider intervals with dyadic rational endpoints, or the intersection of intervals with a given measurable set, or perhaps a more general collection. These more general collections should be in some sense similar to the collection of intervals. We extract two basic properties from the collection of intervals: one captures how intervals behave under set-theoretic operations, and the other characterizes the approximation property. For the first one, note that

- the intersection of any two intervals is an interval, and
- the set difference of two intervals is a finite union of disjoint intervals.

These considerations will lead us to define the notion of a *semi-ring*. The approximation property is captured by the notion of a *sufficient semi-ring*.

On a first reading the reader may concentrate only on the definition of a semi-ring and sufficient semi-ring and the statement of Lemma 2.7.3. Here "sufficient semi-ring" may be replaced by the collection of all intervals or the collection of all dyadic intervals. In fact, the reader has already proven a version of this lemma in Exercise 2.4.3.

A **semi-ring** on a nonempty set X is a collection $\mathcal{R}$ of subsets of X such that

(1) $\mathcal{R}$ is nonempty;

(2) if $A, B \in \mathcal{R}$, then $A \cap B \in \mathcal{R}$;

(3) if $A, B \in \mathcal{R}$, then

$$A \setminus B = \bigsqcup_{j=1}^{n} E_j,$$

where $E_j \in \mathcal{R}$ are disjoint.

A semi-ring must contain the empty set as $\emptyset = A \setminus A$ for any element $A \in \mathcal{R}$. Note that the collection of all intervals (or all bounded intervals) in $\mathbb{R}$ forms a semi-ring. (This is clear since when I and J are any intervals, then $I \cap J$ is empty or an interval, and $I \setminus J$ is empty or a finite union of disjoint intervals, and intervals may be empty.) Similarly, one can verify that the collection of left-closed and right-open intervals is a semi-ring. Also, if $\mathcal{R}$ is a semi-ring of subsets of a nonempty set X and $Y \subset X$, then the collection $\{A : A \cap Y, A \in \mathcal{R}\}$ is a semi-ring. Another example of a semi-ring is given by the collection of all (left-closed, right-open) dyadic intervals. For any set X, the collection of all its subsets is a semi-ring.

One reason semi-rings will be useful is the following proposition. It states that a set that is a countable union of elements of a semi-ring can be written as a countable disjoint union of elements of the semi-ring. (The reader should verify this for intervals.) The process in the proof of Proposition 2.7.1 will be called the process of "disjointifying" the sets A_n and will be useful in later chapters.

Proposition 2.7.1. *Let $\mathcal{R}$ be a semi-ring. If $A = \bigcup_{n=1}^{\infty} A_n$, where $A_n \in \mathcal{R}$, then A can be written as*

$$A = \bigsqcup_{k=1}^{\infty} C_k,$$

where the sets $\{C_k\}$ are disjoint and are in the semi-ring $\mathcal{R}$.

Proof. Define a sequence of sets $\{B_n\}$ by $B_1 = A_1$, and for $n > 1$, $B_n = A_n \setminus (A_1 \cup \ldots \cup A_{n-1})$. The sets $\{B_n\}$ are disjoint and

$$\bigcup_{n \geq 1} A_n = \bigsqcup_{n \geq 1} B_n.$$

However, the B_n's need not be in $\mathcal{R}$. Now we show that each B_n in turn can be written as a finite union of disjoint elements of $\mathcal{R}$. First set $C_1 = B_1$. Next note that as $B_2 = A_2 \setminus A_1$, by property (3) of a semi-ring, B_2 can be written as a finite union of disjoint elements of $\mathcal{R}$; call them $C_2, \ldots, C_{k_1}$.

For $n \geq 3$, observe that

$$A_n \setminus (A_1 \cup \ldots \cup A_{n-1}) = (A_n \setminus A_1) \cap (A_n \setminus A_2) \cap \ldots \cap (A_n \setminus A_{n-1}).$$

2.7. Approximation with Semi-rings

Furthermore, each $(A_n \setminus A_i)$ can be written as a finite union of disjoint sets in $\mathcal{R}$, so

$$(A_n \setminus A_i) = \bigsqcup_{k=1}^{K_{n,i}} E_k^{n,i},$$

for some $E_k^{n,i} \in \mathcal{R}$. But $\mathcal{R}$ is closed under finite intersections and by taking all possible intersections of the sets $E_k^{n,i}$ one can write each set B_n as a finite union of disjoint sets in $\mathcal{R}$. We show this for the case when $n = 3$ and the general case follows by induction. Write

$$A_3 \setminus A_1 = \bigsqcup_{k=1}^{K_{3,1}} E_k^{3,1} \quad \text{and} \quad A_3 \setminus A_2 = \bigsqcup_{\ell=1}^{K_{3,2}} E_\ell^{3,2}.$$

Then let

$$F_{k,\ell} = E_k^{3,1} \cap E_\ell^{3,2},$$

for $k = 1, \ldots, K_{3,1}$, $\ell = 1, \ldots, K_{3,2}$. Then the family of sets $\{F_{k,\ell}\}$, for $k = 1, \ldots, K_{3,1}$, $\ell = 1, \ldots, K_{3,2}$, is disjoint and are all elements of $\mathcal{R}$. It follows that

$$B_3 = (A_3 \setminus A_1) \cap (A_3 \setminus A_2) = \bigsqcup_{k,\ell} F_{k,\ell},$$

a disjoint finite union of elements of the semi-ring $\mathcal{R}$, which we rename $C_{k_1+1}, \ldots, C_{k_2}$.

This yields a collection of disjoint sets $\{C_k\}$ in $\mathcal{R}$ whose union is A. □

Let $(X, \mathcal{S}, \mu)$ be a measure space (in most applications, a canonical Lebesgue measure space). A semi-ring $\mathcal{C}$ of measurable subsets of X of finite measure is said to be a **sufficient semi-ring** for $(X, \mathcal{S}, \mu)$ if it satisfies the following approximation property:

For every $A \in \mathcal{S}$,

$$(2.5) \quad \mu(A) = \inf \left\{ \sum_{j=1}^{\infty} \mu(I_j) : A \subset \bigcup_{j=1}^{\infty} I_j \text{ and } I_j \in \mathcal{C} \text{ for } j \geq 1 \right\}.$$

The definition of a sufficient semi-ring is of interest mainly in the case of nonatomic spaces. The arguments in the atomic case are rather straightforward and will be left to the reader. We have seen, for example, that the collection of all intervals, the collection of all intervals with rational endpoints and the collection of (left-closed, right-open) dyadic intervals are all sufficient semi-rings for $(\mathbb{R}, \mathfrak{L}, \lambda)$. Also, for any measurable set $X \subset \mathbb{R}$, any of these collections of intervals intersected with X forms a sufficient semi-ring for $(X, \mathfrak{L}, \lambda)$ (see Exercise 2). The following lemmas demonstrate why the class of sufficient semi-rings is interesting, and enable us to prove Theorem 3.4.1. The first lemma shows how one can approximate measurable sets up to measure zero with elements from a sufficient semi-ring, but uses countably many of them. The second lemma uses only finitely many elements of the sufficient semi-ring, but in this case the approximation is only "up to ε."

Lemma 2.7.2. *Let $(X, \mathcal{S}, \mu)$ be a measure space with a sufficient semi-ring $\mathcal{C}$. Then for any $A \in \mathcal{S}$ with $\mu(A) < \infty$ there exists a set $H = H(A)$, of the form*

$$H = \bigcap_{n=1}^{\infty} H_n,$$

and such that

(1) $H_1 \supset H_2 \supset \cdots \supset H_n \supset \cdots \supset H \supset A$;

(2) $\mu(H_n) < \infty$;

(3) *each H_n is a countable disjoint union of elements of $\mathcal{C}$; and*

(4) $\mu(H \setminus A) = 0$.

Proof. By the approximation property (2.5) of $\mathcal{C}$, for any $\varepsilon > 0$ there is a set $H(\varepsilon) = \bigcup_{j=1}^{\infty} I_j$, with $I_j \in \mathcal{C}$, such that

$$A \subset H(\varepsilon) \text{ and } \mu(H(\varepsilon) \setminus A) < \varepsilon.$$

Write

$$H_n = H(1) \cap H(1/2) \cap \ldots \cap H(1/n).$$

Since $\mathcal{C}$ is closed under finite intersections, one can verify that each H_n is a countable union of elements of $\mathcal{C}$, and by Proposition 2.7.1 we can write this union as a countable disjoint union of elements of $\mathcal{C}$. Furthermore, by construction the sets H_n are decreasing (i.e.,

2.7. Approximation with Semi-rings

$H_n \subset H_m$ for $n > m$), contain A, and each H_n is of finite measure. Let

$$H = \bigcap_{n=1}^{\infty} H_n.$$

Then H has the required form and

$$\mu(H \setminus A) \leq \mu(H_n \setminus A) < 1/n,$$

for all $n \geq 1$. Therefore $\mu(H \setminus A) = 0$, so $\mu(A) = \mu(H)$. □

We can say that the set H of Lemma 2.7.2 is in $\mathcal{C}_{\sigma\delta}$.

The following lemma will have several applications in later chapters. It is the first of Littlewood's Three Principles, which says that "every (Lebesgue) measurable set is nearly a finite union of intervals."

Lemma 2.7.3. *Let $(X, \mathcal{S}, \mu)$ be a measure space with a sufficient semi-ring $\mathcal{C}$. Let A be a measurable set, $\mu(A) < \infty$, and let $\varepsilon > 0$. Then there exists a set H^* that is a finite union of disjoint elements of $\mathcal{C}$ such that*

$$\mu(A \triangle H^*) < \varepsilon.$$

Proof. As $\mathcal{C}$ is a sufficient semi-ring and $\mu(A) < \infty$, for $\varepsilon > 0$, there exists a set $H[\varepsilon] = \bigcup_{j=1}^{\infty} I_j \supset A$ with $I_j \in \mathcal{C}$ and $\mu(H[\varepsilon]) < \mu(A) + \varepsilon/2$, so $\mu(H[\varepsilon] \setminus A) < \varepsilon/2$. By Proposition 2.5.2,

$$\lim_{n \to \infty} \mu(\bigcup_{j=1}^{n} I_j) = \mu(\bigcup_{j=1}^{\infty} I_j),$$

so there exists $N > 1$ such that

$$0 \leq \mu(H[\varepsilon]) - \mu(\bigcup_{j=1}^{N} I_j) < \varepsilon/2.$$

Let $H^* = \bigcup_{j=1}^{N} I_j$. Then

$$\mu(H^* \triangle A) = \mu(H^* \setminus A) + \mu(A \setminus H^*)$$
$$\leq \mu(H[\varepsilon] \setminus A) + \mu(H[\varepsilon] \setminus H^*)$$
$$< \varepsilon/2 + \varepsilon/2 = \varepsilon.$$

Finally, by Proposition 2.7.1, we may assume that H^* is a disjoint union of elements of $\mathcal{C}$. □

We end this section with another characterization of approximation. First note that in Lemma 2.7.3, the set H^* is a finite union of elements of the semi-ring; so it will be useful to consider collections closed under finite unions. A semi-ring that is closed under finite unions is called a **ring**. In the case when X is of finite measure, most of the rings that we consider also contain X, so they have the additional structure of an algebra (an algebra is just a ring that contains X). The reader should keep the following examples in mind. The typical semi-ring is the collection of all bounded intervals in $\mathbb{R}$ (or all left-closed, right-open bounded intervals in $\mathbb{R}$). The typical ring is the collection of all finite unions of such intervals. The typical algebra is the collection of all finite unions of subintervals of a bounded interval (or all left-closed, right-open subintervals of an interval of the form $[a,b)$). Refer to the exercises at the end of this section for other characterizations of rings and algebras.

Let $(X, \mathcal{S}, \mu)$ be a σ-finite measure space. (In most cases, X will be of finite measure.) A ring $\mathcal{R}$ is said to **generate mod** 0 the σ-algebra $\mathcal{S}$ if the σ-algebra $\sigma(\mathcal{R})$ generated by $\mathcal{R}$ satisfies the property: for all $A \in \mathcal{S}$, $\mu(A) < \infty$, there exists $E \in \sigma(\mathcal{R})$ such that $\mu(A \triangle E) = 0$. Evidently, the ring of finite unions of intervals generates mod 0 the Lebesgue sets. The following lemma gives a characterization of the notion of generation mod 0. A ring that generates mod 0 may also be called a **dense ring**. A **dense algebra** is defined similarly.

Lemma 2.7.4. *Let $(X, \mathcal{S}, \mu)$ be a finite measure space. Let $\mathcal{A}$ be an algebra in X. Then $\mathcal{A}$ generates $\mathcal{S}$ mod 0 if and only if for any $A \in \mathcal{S}$ and any $\varepsilon > 0$ there exists $E \in \mathcal{A}$ such that $\mu(A \triangle E) < \varepsilon$.*

Proof. Suppose $\mathcal{A}$ generates $\mathcal{S}$ mod 0, so $\sigma(\mathcal{A})$ is equal to $\mathcal{S}$ mod 0. Define the collection of sets

$$\mathcal{C} = \{A \in \sigma(\mathcal{A}) : \text{ for } \varepsilon > 0 \text{ there exists } E \in \mathcal{A} \text{ with } \mu(A \triangle E) < \varepsilon\}.$$

Clearly, $\mathcal{C}$ contains $\mathcal{A}$. As $\mu(A^c \triangle E^c) = \mu(A \triangle E)$, it follows that $\mathcal{C}$ is closed under complements. With a bit more work one verifies that $\mathcal{C}$ is closed under countable unions (see Exercise 8). Therefore $\mathcal{C}$ is a σ-algebra. This implies that $\sigma(\mathcal{A})$ is contained in $\mathcal{C}$, completing the proof of this direction.

2.7. Approximation with Semi-rings 45

Now suppose that for any $A \in \mathcal{S}$ and any $\varepsilon > 0$ there exists $E \in \mathcal{A}$ such that $\mu(A \triangle E) < \varepsilon$. For each $n \geq 1$ choose a set $E_n \in \mathcal{A}$ such that
$$\mu(A \triangle E_n) < \frac{\varepsilon}{2^{n+1}}.$$
Then for each $k \geq 1$, $\mu(A \triangle \bigcup_{n=k}^{\infty} E_n) < \frac{\varepsilon}{2^k}$. So
$$\mu(A \triangle \bigcap_{k=1}^{\infty} \bigcup_{n=k}^{\infty} E_n) = 0.$$
Then $F = \bigcap_{k=1}^{\infty} \bigcup_{n=k}^{\infty} E_n \in \sigma(\mathcal{A})$ and $\mu(A \triangle F) = 0$. □

Corollary 2.7.5. *Let $(X, \mathcal{S}, \mu)$ be a finite measure space. Let $\mathcal{C}$ be a semi-ring in $\mathcal{S}$ containing X. Then $\mathcal{C}$ is a sufficient semi-ring only if the algebra consisting of all finite unions of elements of $\mathcal{C}$ is dense.*

We conclude with some alternative notation to express the notions of approximation. We think of two measurable sets A and B as being "ε-close" (for some $\varepsilon > 0$) when $\lambda(A \triangle B) < \varepsilon$. We state some properties of this "distance."

Proposition 2.7.6. *Let $(X, \mathcal{S}, \mu)$ be a probability measure space. For any measurable sets A and B, define*
$$D(A, B) = \mu(A \triangle B).$$
Then

(1) $D(A, B) = 0$ *if and only if* $A = B$ mod μ;
(2) $D(A, B) = D(B, A)$;
(3) $D(A, B) \leq D(A, C) + D(C, B)$, *for any measurable set C;*
(4) *if* $D(A, B) < \varepsilon$, *then* $|\mu(A) - \mu(B)| < \varepsilon$.

It follows that D is a pseudo-metric; it is not a metric as it may happen that $A \neq B$ but $D(A, B) = 0$ (in fact, for any null set N, $D(N, \emptyset) = 0$); but D obeys the other properties of a metric. Also, using this notation it follows that if $\mathcal{C}$ is a sufficient ring, then for all measurable sets A and any $\varepsilon > 0$ there exists $C \in \mathcal{C}$ so that $D(B, C) < \varepsilon$. When working with probability spaces, most rings that one studies already contain X. A ring that contains X is an algebra (see Exercise 5). An important algebra in $[0, 1]$ is the collection of all finite unions of dyadic intervals.

Exercises

(1) Show that if $\mathcal{C}$ is a semi-ring of subsets of a nonempty set X and $\emptyset \neq Y \subset X$, then the collection $\{A \cap Y : A \in \mathcal{C}\}$, if nonempty, is a semi-ring of subsets of Y. Show a similar property for the case when $\mathcal{C}$ is a ring.

(2) Let $(X, \mathfrak{L}, \lambda)$ be a canonical Lebesgue measure space and $\mathcal{C}$ a sufficient semi-ring. Show that for any nonempty measurable set $X_0 \subset X$, $\mathcal{C} \cap X_0 = \{C \cap X_0 : C \in \mathcal{C}\}$ is a sufficient semi-ring for $(X_0, \mathfrak{L}(X_0), \lambda)$.

(3) Show that a ring is closed under symmetric differences and finite intersections.

(4) Let $\mathcal{R}$ be a nonempty collection of subsets of a set X such that for all $A, B \in \mathcal{R}$, $A \cup B \in \mathcal{R}$ and $A \setminus B \in \mathcal{R}$. Show that $\mathcal{R}$ is a ring.

(5) Let $\mathcal{A}$ be a nonempty collection of subsets of a set X. Show that $\mathcal{A}$ is an algebra if and only if it is closed under complements and finite intersections. Give an example of a ring that is not an algebra. Give an example of an algebra that is not a σ-algebra.

(6) Show that if $\mathcal{C}$ is a collection of sets, then the intersection of all rings containing $\mathcal{C}$ is a ring, called the **ring generated by** $\mathcal{C}$ and denoted by $r(\mathcal{C})$.

(7) Show that if $\mathcal{R}$ is a semi-ring, then the ring $r(\mathcal{R})$ generated by $\mathcal{R}$ is obtained by taking all finite unions of disjoint elements from $\mathcal{R}$.

(8) Let $\mathcal{C}$ be defined as in the proof of Lemma 2.7.4. Show that $\mathcal{C}$ is closed under countable unions.

(9) Prove Proposition 2.7.6.

(10) Prove Corollary 2.7.5.

* (11) Let X be a Lebesgue measurable set in $\mathbb{R}$ and let d be the metric defined on the set $\mathfrak{L}(X)/\sim$ in Exercise 2.5.13. Let $\mathcal{C}$ be a sufficient semi-ring in $(X, \mathcal{L}(X), \lambda)$ and let $r(\mathcal{C})$ be the ring generated by $\mathcal{C}$. Show that in the metric space $(\mathfrak{L}(X)/\sim, d)$ the collection $\{[H] : H \in \mathcal{R}\}$ is a dense set.

(12) Let X be a nonempty set and let $\mathcal{R}$ be a ring of subsets of X. Define a σ-ring to be a ring that is closed under countable unions and countable intersections. Show that the smallest σ-ring containing $\mathcal{R}$ is equal to the smallest monotone class containing $\mathcal{R}$.

2.8. Measures from Outer Measures

Lebesgue measure can be extended in a natural way from $\mathbb{R}$ to $\mathbb{R}^d$ for any integer $d > 1$. Lebesgue measure for $d = 2$ should generalize the notion of area, and for $d = 3$ the notion of volume. The idea is to replace intervals, in the definition of Lebesgue outer measure in the line, by Cartesian products of intervals in the definition of outer measure in $\mathbb{R}^d$. Rather than developing d-dimensional Lebesgue measure in a way similar to our construction of Lebesgue measure, we instead introduce a method that works in a more general context. This method is based on Carathéodory's condition for measurability.

Before presenting the main definitions we introduce some notation. A **set function** is a function μ, defined on some collection $\mathcal{C}$ of subsets of a nonempty set X with $\emptyset \in \mathcal{C}$, such that μ has values in $[-\infty, \infty]$ (all the set functions we consider will have values in $[0, \infty]$). A set function $\mu : \mathcal{C} \to [0, \infty]$ is said to be **finitely additive** on $\mathcal{C}$ if $\mu^*(\emptyset) = 0$ and for any disjoint sets $A_i, 1 \leq i \leq n$, in $\mathcal{C}$ such that $\bigsqcup_{i=1}^{n} A_i \in \mathcal{C}$, it is the case that

$$\mu^*(\bigsqcup_{i=1}^{n} A_i) = \sum_{i=1}^{n} \mu^*(A_i).$$

A set function $\mu : \mathcal{C} \to [0, \infty]$ is **countably additive** on $\mathcal{C}$ if $\mu^*(\emptyset) = 0$ and for any sequence of disjoint sets $A_i, i \geq 1$, in $\mathcal{C}$ such that $\bigsqcup_{i=1}^{\infty} A_i \in \mathcal{C}$, it is the case that

$$\mu^*(\bigsqcup_{i=1}^{\infty} A_i) = \sum_{i=1}^{\infty} \mu^*(A_i).$$

Certainly, Lebesgue measure on the semi-ring of bounded subintervals of $\mathbb{R}$ is a countably additive set function. A set function that is finitely additive but not countably additive is given in Exercise 1. We will mainly be interested in finite additivity as an intermediate property that is verified before eventually showing that the set function under consideration is countably additive.

Given a nonempty set X, an **outer measure** μ^* is a set function defined on all subsets of X and with values in $[0, \infty]$ satisfying the following properties:

(1) $\mu^*(\emptyset) = 0$;
(2) μ^* is **monotone**: if $A \subset B$, then $\mu^*(A) \leq \mu^*(B)$;
(3) μ^* is **countably subadditive**: $\mu^*(\bigcup_{i=1}^\infty A_i) \leq \sum_{i=1}^\infty \mu^*(A_i)$.

The first example we have of an outer measure is Lebesgue outer measure. Any measure defined on all subsets of a finite set is also an outer measure.

As we have seen, one of the most important properties to prove in our development of Lebesgue measure was countable additivity. Part of the proof involved choosing the right class of sets on which the outer measure is a measure, in this case the σ-algebra of Lebesgue measurable sets. Carathéodory introduced a remarkable property that does precisely this for any outer measure. A set A is said to be μ^*-**measurable** if for any set C,

$$\mu^*(C) = \mu^*(A \cap C) + \mu^*(A^c \cap C).$$

This is the same as saying that a set A is μ^*-measurable if and only if for any set C_1 contained in A and any set C_2 contained in its complement A^c,

(2.6) $$\mu^*(C_1 \sqcup C_2) = \mu^*(C_1) + \mu^*(C_2).$$

(This follows by letting $C = C_1 \cup C_2$ in the definition.) In other words, A is μ^*-measurable if and only if A partitions the sets in X so that μ^* is additive on the disjoint union of a set contained in A with a set contained in its complement. While this characterization gives an idea of the condition, its true merit lies in the fact that Theorem 2.8.1 holds.

Question. Show that any set A with $\mu^*(A) = 0$ is μ^*-measurable.

Theorem 2.8.1. *Let X be a nonempty set and let μ^* be an outer measure on X. If $\mathcal{S}(\mu^*)$ denotes the μ^*-measurable sets in X, then $\mathcal{S}(\mu^*)$ is a σ-algebra and μ^* restricted to $\mathcal{S}(\mu^*)$ is a measure.*

2.8. Measures from Outer Measures

Proof. It follows immediately from the definition that $\mathcal{S}(\mu^*)$ contains the empty set and is closed under complements. We now show that $\mathcal{S}(\mu^*)$ is closed under finite unions. First we show that this is the case for finite unions of disjoint sets. So let $A, B \in \mathcal{S}(\mu^*)$ with $A \cap B = \emptyset$. Let $C_1 \subset A \sqcup B$ and $C_2 \subset (A \sqcup B)^c$. Write $C_{1,1} = C_1 \cap A$ and $C_{1,2} = C_1 \cap B$. As A is μ^*-measurable, $C_{1,1} \subset A$, and $C_{1,2} \subset A^c$,

$$\mu^*(C_1) = \mu^*(C_{1,1} \sqcup C_{1,2}) = \mu^*(C_{1,1}) + \mu^*(C_{1,2}).$$

Similarly, using that B is μ^*-measurable,

$$\mu^*(C_{1,2} \sqcup C_2) = \mu^*(C_{1,2}) + \mu^*(C_2).$$

Therefore, again using that A is μ^*-measurable and then applying the previous results,

$$\mu^*(C_1 \sqcup C_2) = \mu^*(C_{1,1} \sqcup (C_{1,2} \sqcup C_2)) = \mu^*(C_{1,1}) + \mu^*(C_{1,2} \sqcup C_2)$$
$$= \mu^*(C_{1,1}) + \mu^*(C_{1,2}) + \mu^*(C_2)$$
$$= \mu^*(C_{1,1} \sqcup C_{1,2}) + \mu^*(C_2)$$
$$= \mu^*(C_1) + \mu^*(C_2).$$

Therefore, $A \sqcup B$ is in $\mathcal{S}(\mu^*)$.

Next we show in a similar way that if $A, B \in \mathcal{S}(\mu^*)$, then $A \cap B \in \mathcal{S}(\mu^*)$. In fact, let $C_1 \subset A \cap B$ and $C_2 \subset A^c \cup B^c$. Let $C_{2,1} = C_2 \cap A$ and $C_{2,2} = C_2 \cap A^c$. Then

$$\mu^*(C_1 \sqcup C_2) = \mu^*((C_1 \sqcup C_{2,1}) \sqcup C_{2,2}) = \mu^*(C_1 \sqcup C_{2,1}) + \mu^*(C_{2,2})$$
$$= \mu^*(C_1) + \mu^*(C_{2,1}) + \mu^*(C_{2,2})$$
$$= \mu^*(C_1) + \mu^*(C_{2,1} \sqcup C_{2,2})$$
$$= \mu^*(C_1) + \mu^*(C_2).$$

So $\mathcal{S}(\mu^*)$ is closed under finite intersections. Since $\mathcal{S}(\mu^*)$ is closed under complements, finite disjoint unions and finite intersections, then it is closed under finite unions.

Before considering countably many sets we observe that μ^* is finitely additive on μ^*-measurable sets. In fact if A and B are μ^*-measurable and disjoint, as $A \subset A$ and $B \subset A^c$, then $\mu^*(A \sqcup B) = \mu^*(A) + \mu^*(B)$.

It remains to show that $\mathcal{S}(\mu^*)$ is closed under countable unions and that μ^* is countably additive. Let $\{A_i\}_{i \geq 1}$ be a sequence of sets in $\mathcal{S}(\mu^*)$. We wish to verify that $\bigcup_i A_i$ is μ^*-measurable. Since we now know that $\mathcal{S}(\mu^*)$ is an algebra, hence a semi-ring, Proposition 2.7.1 implies that we may assume that the sets A_i are disjoint. Let

$$C_1 \subset \bigsqcup_{i=1}^{\infty} A_i, \quad C_2 \subset \left(\bigsqcup_{i=1}^{\infty} A_i\right)^c,$$

and set

$$C_{1,n} = C_1 \cap \bigsqcup_{i=1}^{n} A_i.$$

As $C_2 \subset (\bigsqcup_{i=1}^{n} A_i)^c$, $\mu^*(C_{1,n} \sqcup C_2) = \mu^*(C_{1,n}) + \mu^*(C_2)$. Therefore,

$$\mu^*(C_1 \sqcup C_2) \geq \lim_{n \to \infty} \mu^*(C_{1,n} \sqcup C_2)$$
$$= \lim_{n \to \infty} \mu^*(C_{1,n}) + \mu^*(C_2).$$

But,

$$\mu^*(C_1) = \mu^*(\bigsqcup_{i=1}^{\infty} A_i \cap C_1) \leq \sum_{i=1}^{\infty} \mu^*(A_i \cap C_1)$$
$$= \lim_{n \to \infty} \sum_{i=1}^{n} \mu^*(A_i \cap C_1)$$
$$= \lim_{n \to \infty} \mu^*(\bigsqcup_{i=1}^{n} A_i \cap C_1)$$
$$= \lim_{n \to \infty} \mu^*(C_{1,n}).$$

Therefore, $\mu^*(C_1 \sqcup C_2) \geq \mu^*(C_1) + \mu^*(C_2)$. This shows that $\bigsqcup_{i=1}^{\infty} A_i \in \mathcal{S}(\mu^*)$.

The fact that μ^* is countably additive follows an argument we have already seen:

$$\mu^*\left(\bigsqcup_{i=1}^{\infty} A_i\right) \geq \lim_{n \to \infty} \mu^*\left(\bigsqcup_{i=1}^{n} A_i\right)$$
$$= \lim_{n \to \infty} \sum_{i=1}^{n} \mu^*(A_i) = \sum_{i=1}^{\infty} \mu^*(A_i).$$

□

2.8. Measures from Outer Measures

Application: A Construction of Lebesgue measure on $\mathbb{R}^d$.

Recall that the **Cartesian product** of d sets $A_1, \ldots, A_d$ is defined to be
$$A_1 \times \ldots \times A_d = \{(x_1, \ldots, x_d) : x_i \in A_i, \text{ for } i = 1, \ldots, d\}.$$

A d**-rectangle** or **rectangle** I in $\mathbb{R}^d$ is defined to be a set of the form
$$I = I_1 \times \ldots \times I_d,$$
where I_j, $j \in \{1, \ldots, d\}$, are bounded intervals in $\mathbb{R}$.

Define a set function on the collection of d-rectangles by
$$|I|_{(d)} = |I_1| \times \ldots \times |I_d|,$$
called the d**-volume**. Clearly, for $d = 1$ this corresponds to the length on an interval, for $d = 2$ to the area of a square and for $d = 3$ to the volume of a 3-rectangle.

Define the d**-dimensional Lebesgue outer measure** in $\mathbb{R}^d$ by (2.7)
$$\lambda_{(d)}^*(A) = \inf\{\sum_{j=1}^\infty |I_j|_d : A \subset \bigcup_{j=1}^\infty I_j, \text{where } I_j \text{ are } d\text{-rectangles}\}.$$

By Theorem 2.8.1 this outer measure is a measure when restricted to the σ-algebra $S(\lambda_{(d)}^*)$. It remains to show that d-rectangles are $\lambda_{(d)}^*$-measurable. This will imply that the Borel sets in $\mathbb{R}^d$ are $\lambda_{(d)}^*$-measurable, so $\lambda_{(d)}^*$ restricted to the Borel sets is a measure, giving a construction of Lebesgue measure on the Borel σ-algebra of $\mathbb{R}^d$. For a proof that the Borel sets are $\lambda_{(1)}^*$-measurable in the 1-dimensional case, that easily extends to the d-dimensional case. For a complete proof the reader may consult [**14**, Proposition 1.3.5]. Another construction is given later as an application of the following theorem.

One can use the notion of outer measure to construct more general measures by extending a finitely additive set function defined on a semi-ring of sets. When working with semi-rings we shall make the assumption that the set X can be written as a countable union of elements of the semi-ring $\mathcal{R}$. While some theorems can be proved without this assumption, all the cases we shall be interested in satisfy the assumption.

2. Lebesgue Measure

Theorem 2.8.2 (Carathéodory Extension Theorem). *Let X be a nonempty set and let $\mathcal{R}$ be a semi-ring of subsets of X such that X can be written as a countable union of elements of $\mathcal{R}$. Let μ be a countably additive set function on $\mathcal{R}$. Then μ extends to a measure defined on the σ-algebra $\sigma(\mathcal{R})$ generated by $\mathcal{R}$.*

Proof. We first outline the structure of the proof. We start by constructing an outer measure μ^* from μ. Then we show that the elements of $\mathcal{R}$ are μ^*-measurable, which implies that the σ-algebra generated by $\mathcal{R}$ is contained in $\mathcal{S}(\mu^*)$. Thus μ^* defines a measure on $\sigma(\mathcal{R})$, which we also denote by μ.

Start by defining μ^* for any set $A \subset X$, by

$$\mu^*(A) = \inf\left\{\sum_{i=1}^{\infty} \mu(E_i) : A \subset \bigcup_{i=1}^{\infty} E_i, E_i \in \mathcal{R}\right\}.$$

We now show that μ^* is countably subadditive. This is similar to the corresponding proof for the case of Lebesgue outer measure.

Let $A = \bigcup_{i=1}^{\infty} A_i$. We may assume that $\mu(A_i) < \infty$ for $i \in \mathbb{N}$. Let $\varepsilon > 0$. Then for each i there exist sets $E_{i,j} \in \mathcal{R}, j \geq 1$, such that $A_i \subset \bigcup_{j=1}^{\infty} E_{i,j}$, and

$$\sum_{j=1}^{\infty} \mu(E_{i,j}) < \mu^*(A_i) + \frac{\varepsilon}{2^i}.$$

Then

$$\sum_{i=1}^{\infty} \sum_{j=1}^{\infty} \mu(E_{i,j}) \leq \sum_{i=1}^{\infty} \mu^*(A_i) + \varepsilon.$$

As $A \subset \bigcup_i \bigcup_j E_{i,j}$, $\mu^*(A) \leq \sum_{i=1}^{\infty} \mu^*(A_i) + \varepsilon$. Letting $\varepsilon \to 0$ completes the argument. As the other properties of an outer measure are easy to verify we conclude that μ^* is an outer measure.

Next, we show that the elements of $\mathcal{R}$ are μ^*-measurable. Let $E \in \mathcal{R}$ and suppose that $C_1 \subset E$ and $C_2 \subset E^c$. If $\mu^*(C_1) = \infty$ or $\mu^*(C_2) = \infty$, then $\mu^*(C_1 \sqcup C_2) = \infty$, and (2.6) is trivially verified. So suppose that $\mu^*(C_1 \sqcup C_2) < \infty$. For $\varepsilon > 0$ there exist $K_i \in \mathcal{R}$ such

2.8. Measures from Outer Measures

that $C_1 \sqcup C_2 \subset \bigsqcup_{i=1}^{\infty} K_i$ and

$$\mu^*(C_1 \sqcup C_2) > \left[\sum_{i=1}^{\infty} \mu(K_i)\right] - \varepsilon.$$

We can write $K_i = (K_i \cap E) \sqcup (K_i \cap E^c)$ and, as $\mathcal{R}$ is a semi-ring,

$$K_i \cap E^c = K_i \setminus E = \bigsqcup_{j=1}^{n_i} F_{i,j},$$

for some $F_{i,j} \in \mathcal{R}$. Thus,

$$\mu^*(C_1 \sqcup C_2) > \left[\sum_{i=1}^{\infty} \mu(K_i \cap E \sqcup \bigsqcup_{j=1}^{n_i} F_{i,j})\right] - \varepsilon$$

$$= \left[\sum_{i=1}^{\infty} \mu(K_i \cap E)\right] + \left[\sum_{i=1}^{\infty} \sum_{j=1}^{n_i} \mu(F_{i,j})\right] - \varepsilon$$

$$\geq \mu^*(C_1) + \mu^*(C_2) - \varepsilon.$$

This shows that the elements of $\mathcal{R}$ are μ^*-measurable. Therefore $\sigma(\mathcal{R})$ is contained in $\mathcal{S}(\mu^*)$.

We observe that μ^* restricted to $\mathcal{R}$ agrees with μ. Clearly, as $E \in \mathcal{R}$ covers itself, $\mu^*(E) \leq \mu(E)$. Then Exercise 2 implies that $\mu(E) \leq \mu^*(E)$. □

The hypothesis of Theorem 2.8.2 can be relaxed slightly when combined with Exercise 5, whose proof uses ideas already discussed in this section.

Sometimes the Carathéodory extension theorem is stated in the following form. The first part is an immediate consequence of Theorem 2.8.2.

Theorem 2.8.3. *Let X be a nonempty set and let $\mathcal{A}$ be an algebra of subsets of X. If μ is a countably additive set function on $\mathcal{A}$, then it extends to a measure on the σ-algebra generated by $\mathcal{A}$. If μ is σ-finite, then the extension is unique.*

Proof. The existence of the extension follows from Theorem 2.8.2. We show uniqueness of the extension. This is a standard argument for uniqueness and uses the monotone class theorem. We show this in the case when μ is finite and leave the general case to the reader. (For the general case the reader should consider $X = \bigcup_{i=1}^{\infty} K_i$, with $\mu(K_i) < \infty, K_i \in \mathcal{A}$.) Let ν be any other measure on $\sigma(\mathcal{A})$ that agrees with μ on all elements of $\mathcal{A}$. Form the set

$$\mathcal{C} = \{A \in \sigma(\mathcal{A}) : \mu(A) = \nu(A)\}.$$

Evidently, $\mathcal{C}$ contains $\mathcal{A}$. If we show that $\mathcal{C}$ is a monotone class, the monotone class theorem implies that $\mathcal{C}$ contains $\sigma(\mathcal{A})$, completing the proof. But the fact that $\mathcal{C}$ is closed under monotone unions and intersections, and is a monotone class, follows from Proposition 2.5.2. □

We now discuss briefly that it can be shown that a countably additive set function on a semi-ring extends to the generated ring. We will typically not need this result, as we have shown in Theorem 2.8.2 a countably additive set function only needs to be defined on a semi-ring to have an extension to the generated σ-algebra, but is used for uniqueness of the extension as seen below. Recall that $r(\mathcal{R})$ denotes the ring generated by a semi-ring $\mathcal{R}$ and consists of all finite unions of elements of $\mathcal{R}$ (Exercise 2.7.4). We simply give a brief outline of the proof; for a complete proof see [**68**, Theorem 3.5].

Proposition 2.8.4. *Let $\mathcal{R}$ be a semi-ring on a set X. If μ is a countably additive set function on $\mathcal{R}$, then it has a unique extension to $r(\mathcal{R})$.*

Proof. Let μ also denote the extension of μ. It has a natural definition on $r(\mathcal{R})$. Let $A \in r(\mathcal{R})$. Then A can be written as $A = \bigsqcup_{i=1}^{n} K_i$ for some $K_i \in \mathcal{R}$ and $n > 0$. Then define μ by

$$\mu(A) = \sum_{i=1}^{n} \mu(K_i).$$

It remains to show that μ is well defined and countably additive on $r(\mathcal{R})$. □

As a consequence we obtain a more general result on uniqueness of the extension than the one in Theorem 2.8.3.

2.8. Measures from Outer Measures

Lemma 2.8.5. *Let X be a nonempty set and let $\mathcal{R}$ be a semi-ring of subsets of X such that X can be written as a countable union of elements of $\mathcal{R}$. Let μ be a countably additive set function on $\mathcal{R}$ that is finite on $\mathcal{R}$. Then any measure ν on $\sigma(\mathcal{R})$ that agrees with μ on $\mathcal{R}$ must agree with μ on $\sigma(\mathcal{R})$.*

Proof. To apply the monotone class theorem as in the proof of Theorem 2.8.3 we need $\mathcal{R}$ to be an algebra. As $\mathcal{R}$ is not necessarily an algebra our argument consists of techniques to reduce it to that case.

First we note that by Proposition 2.8.4 ν defined on $\mathcal{R}$ has a unique extension to the ring generated by $\mathcal{R}$, so ν must agree with μ on $r(\mathcal{R})$. From the assumption on $\mathcal{R}$ there exists a sequence of sets $X_n \in \mathcal{R}$, $\mu(X_n) < \infty$, such that $X_1 \subset X_2 \subset \cdots \subset X_n \subset \cdots$ and for any $A \in \sigma(\mathcal{R})$, $\nu(A) = \lim_{n\to\infty} \nu(A \cap X_n)$. Then, for each $n > 0$, ν is unique on the algebra in X_n generated by $\mathcal{R}$. Then the monotone class theorem argument applies. $\square$

The following observation is elementary but useful.

Lemma 2.8.6. *Let X be a nonempty set and let $\mathcal{R}$ be a semi-ring of subsets of X such that X can be written as a countable union of elements of $\mathcal{R}$. Let μ be a countably additive set function on $\mathcal{R}$ that is finite on elements of $\mathcal{R}$. Then $\mathcal{R}$ is a sufficient semi-ring for the extension of μ as in Theorem 2.8.2.*

Before discussing some applications we prove a lemma that has some useful conditions to show that a finitely additive set function on a ring is countably additive.

Lemma 2.8.7. *Let $\mathcal{R}$ be a ring of subsets of X and let μ be a finitely additive measure on $\mathcal{R}$.*

(1) *If for all $A_i, A \in \mathcal{R}$ with $A_1 \subset A_2 \subset \cdots$ and $A = \bigcup_{i=1}^{\infty} A_i$ we have $\lim_{i\to\infty} \mu(A_i) = \mu(A)$, then μ is countably additive.*

(2) *If μ is finite on $\mathcal{R}$ and for all $A_i, A \in \mathcal{R}$ with $A_1 \supset A_2 \supset \cdots$ and $\bigcap_{i=1}^{\infty} A_i = \emptyset$ we have $\lim_{i\to\infty} \mu(A_i) = 0$, then μ is countably additive.*

Proof. For part (1), if $\{B_i\}$ is any disjoint sequence of elements of $\mathcal{R}$ with $A = \bigsqcup_{i=1}^{\infty} B_i \in \mathcal{R}$, then if $A_n = \bigsqcup_{i=1}^{n} B_i$ both A_i and A satisfy the hypotheses of the lemma and

$$\mu(\bigsqcup_{i=1}^{\infty} B_i) = \lim_{n\to\infty} \mu(\bigsqcup_{i=1}^{n} B_i) = \lim_{n\to\infty} \sum_{i=1}^{n} \mu(B_i) = \sum_{i=1}^{\infty} \mu(B_i).$$

Finally, we show that, under the hypothesis of part (2), the hypothesis of part (1) holds. So let $\{A_i\}$ be an increasing sequence of sets in $\mathcal{R}$ and $A = \bigcup_{i=1}^{\infty} A_i$. Set $B_i = A \setminus A_i$. Then $\{B_i\}$ is a decreasing sequence of sets and $\bigcap_{i=1}^{\infty} B_i = \emptyset$. Then $\mu(B_i) \to 0$, or $\mu(A \setminus A_i) \to 0$. As $\mu(A) < \infty$, this implies $\mu(A_i) \to \mu(A)$. □

Application: Another Construction of Lebesgue measure on $\mathbb{R}^d$.

The reader is asked to prove the following lemma. (A similar lemma is shown in Lemma 4.9.1.)

Lemma 2.8.8. *The collection $\mathcal{R}_d$ of all d-rectangles is a semi-ring.*

First one shows that the set function $|\cdot|_{(d)}$ defined earlier is finitely additive on the semi-ring of rectangles. The proof for $d = 1$ is similar to the proof of Lemma 2.1.2. The proof for $d > 1$ is reduced to the 1-dimensional case by the appropriate decomposition of the d-rectangle. The reader is asked to do this in the exercises (for the details in the 2-dimensional case refer to [**68**, Section 3.4]).

The next step is to use a compactness argument to show that $|\cdot|_{(d)}$ is countably additive on the semi-ring of rectangles. A proof of this is in [**68**, Section 3.4]. Then by Theorem 2.8.2 $|\cdot|_{(d)}$ extends to a unique measure on the Borel sets of $\mathbb{R}^d$.

Exercises

(1) Let $X = \mathbb{N}$ and let $\mathcal{C}$ consist of all subsets of $\mathbb{N}$. Define a set function on $\mathcal{C}$ by $\mu(C) = \sum_{i \in C} \frac{1}{2^i}$ when C is a finite set and $\mu(C) = \infty$ when C is an infinite set. Show that μ is a finitely additive set function that is not countably additive.

(2) Let μ be a countably additive set function on a semi-ring $\mathcal{R}$. Show that if $A, K_i \in \mathcal{R}$ and $A \subset \bigcup_i K_i$, then $\mu(A) \leq \sum_{i=1}^{\infty} \mu(K_i)$.

(3) In Lemma 2.8.7, part (2), give a direct proof that μ is countably additive without reducing it to part (1).

2.8. Measures from Outer Measures

(4) Show that $r(\mathcal{R})$, the ring generated by a semi-ring $\mathcal{R}$, consists of all finite unions of elements of $\mathcal{R}$.

(5) Let X be a nonempty set and let $\mathcal{R}$ be a semi-ring on X. Let μ be a set function on $\mathcal{R}$. Show that μ is countably additive if and only if it is additive and countably subadditive.

(6) Show that if $\mathcal{R}$ is a ring on a set X, then the collection $\mathcal{R} \cup \{X \setminus A : A \in \mathcal{R}\}$ is an algebra on X.

(7) Complete the details in the proof of Lemma 2.8.5.

* (8) Show that the d-volume $|\cdot|_d$ is finitely additive on the semi-ring of rectangles $\mathcal{R}_d$.

* (9) Show that the Borel sets in $\mathbb{R}$ are $\lambda_{(1)}^*$-measurable where $\lambda_{(1)}^*$ is defined as in (2.7). Generalize this to $d > 1$.

Project A. We will extend the notion of outer measure on the line. A similar extension could be done on the plane. For any real numbers $0 < t \leq 1$ and $\delta > 0$ and any set A define

$$\mathcal{H}_\delta^t(A) = \inf\{\sum_j |I_j|^t : A \subset \bigcup_j I_j \text{ and } |I_j| < \delta\}.$$

The t-dimensional Hausdorff measure of a set A is defined by

$$\mathcal{H}^t(A) = \lim_{\delta \to 0} \mathcal{H}_\delta^t(A).$$

Observe that if $\delta_1 > \delta_2$, then $\mathcal{H}_{\delta_1}^t(A) \leq \mathcal{H}_{\delta_2}^t(A)$. Deduce that the limit above exists though it may be infinite. Show that $\mathcal{H}^t(A)$ satisfies the same properties as we showed Lebesgue outer measure satisfied.

The Hausdorff dimension of a set A is defined by

$$\dim_H(A) = \inf\{t \geq 0 : \mathcal{H}^t(A) = 0\}.$$

Show that $d = \dim_H(A)$ is such that $\mathcal{H}^t(A) = \infty$ for $0 \leq t < d$ and $\mathcal{H}^t(A) = 0$ for $t > d$. Compute the Hausdorff dimension for some examples. In particular show that the Cantor middle-thirds set has Hausdorff dimension $t = \frac{\log 2}{\log 3}$ and its $\frac{\log 2}{\log 3}$-Hausdorff measure is 1.

Open Question A. This is an open question due to Erdös. Let $f : \mathbb{R} \to \mathbb{R}$ be the affine map defined by $f(x) = ax + b$ for any $a, b \in \mathbb{R}, a \neq 0$. Two sets A and B in $\mathbb{R}$ are said to be similar if $f(A) = B$ for some map f. Let A be a countable set in $\mathbb{R}$ such that 0 is its only accumulation point. Is it the case that there exists a set $E \subset \mathbb{R}$ with $\lambda(E) > 0$, such that no subset of E is similar to A? This question remains open but special cases are known. a) Search the literature to find out the special solutions to this question by Eigen and Falconer. b) Search the literature to find out what is currently known about this question. c) Write a paper describing in detail the partial solutions of Eigen and/or Falconer. d) Write a paper describing the state of the art on this question. Has this question been answered for the sequence $\{1/2^n : n > 0\}$? e) Solve the problem.

Chapter 3

Recurrence and Ergodicity

This chapter introduces two basic notions for dynamical systems: recurrence and ergodicity. As we saw in Chapter 1, at its most basic level, an abstract dynamical system with discrete time consists of a set X and a map or transformation T defined on the set X and with values in X. We think of X as the set of all possible *states* of the system and of T as the law of time evolution of the system. If the state of the system is x_0 at a certain moment in time, after one unit of time the state of the system will be $T(x_0)$, after two units of time it will be $T(T(x_0))$ (which we denote by $T^2(x_0)$), etc. We are first interested in studying what happens to states and to sets of states as the system evolves through time.

We shall study dynamical properties from a measurable or probabilistic point of view and thus impose a measurable structure on the set X and the map T. The basic structure on X is that of a measure space. While one can define many of the notions and prove some of the theorems in the setting of a general measure space (finite or σ-finite), for all the examples of interest and for many of the theorems with richer structure, one needs to assume some additional structure on X, such as that of a canonical Lebesgue measure space. We require T to be a measurable transformation, which we further specify

60   3. Recurrence and Ergodicity

to be measure-preserving. We start, however, in Section 3.1, with an example that can be used to informally introduce some of the main ideas of this chapter and of Chapter 6.

Topological dynamics has been intimately connected with ergodic theory since the origins of both subjects, and we shall present some concepts from topological dynamics, such as minimality and topological transitivity.

3.1. An Example: The Baker's Transformation

We shall use the baker's transformation to introduce some important concepts in ergodic theory. These concepts are treated in this section in an informal manner and are studied in more detail in later sections.

When kneading dough, a traditional baker uses the following process: stretch the dough, then fold it and perform a quarter turn before repeating the stretch, fold and quarter turn process. The quarter turn is important when mixing a 3-dimensional piece of dough, but can be omitted in the 2-dimensional model. The 2-dimensional case already exhibits all the dynamical behaviors we are interested in but, before studying that, let us consider the 3-dimensional process without the quarter turn applied to the cube of Figure 3.1.

Figure 3.1. Kneading 3-d dough

Consider a right-handed 3-d coordinate system oriented so that the shaded face (marked L, R) of the cube in Figure 3.1 is on the y, z-plane (the plane $x = 0$), with the bottom left corner of the part marked L at $(0, 0, 0)$, the bottom right corner of the part marked R at the point $(0, 1, 0)$, the top right corner of the part marked R at the point $(0, 1, 1)$, the top left corner of the part marked L at $(0, 0, 1)$ and the back corner of the cube that is not showing at the

3.1. An Example: The Baker's Transformation 61

point $(-1, 0, 0)$, so most of the cube, except for the shaded face, is on the negative x-axis. As the reader may verify, under this modified process each vertical cross section (i.e., a point in the plane $x = c$ ($-1 \leq c \leq 0$)) remains invariant. So for example, points on the shaded face ($x = 0$) remain on the shaded face after an application of one iteration of the process; in fact, a point on the plane $x = b$ never reaches the plane $x = c$ for $b \neq c$. This clearly would not be a desirable process for mixing dough, as points on one half of the cube (points with $0 \geq x > -1/2$) never reach the other half (points with $-1/2 \geq x \geq -1$). However, as we shall see in Section 6.6 and explain informally in this section, there is mixing on each cross section of the form $x = a, 0 \geq a \geq -1$. As each cross section of the form $x = a, 0 \geq a \geq -1$, exhibits all the dynamical properties we wish to study, we shall confine our analysis to the 2-dimensional case from now on.

The modified 3-dimensional example demonstrates an important concept: each vertical cross section is mapped to itself, and so there are positive volume subsets of the cube that are mapped to themselves. A subset A of the cube is said to be *invariant* if the iterate (under the process) of every point (x, y, z) in A remains in the set A. Evidently, cross sections with $x = c$ are invariant, and so in particular half the cube (the subset constrained by $0 \geq x \geq -1/2$) is invariant. An invariant set A gives rise to a new dynamical system as one can consider the transformation restricted to the invariant set as a dynamical system in its own right. It can be shown, however, that the original process, including the quarter turn, on the cube does not have any invariant sets of positive volume.

To describe the 2-dimensional process, we look more carefully at the action of the modified 3-dimensional process when applied to 2-dimensional cross sections of the form $x = a, 0 \geq a \geq -1$, which are squares of side length 1. We concentrate on the cross section at $x = 0$ and have shaded each half of it with different intensity to better visualize the process. One full cycle of the kneading process now consists of a stretch and then a fold; stretching produces a figure as in the right; then there is a folding that occurs to return to the original shape.

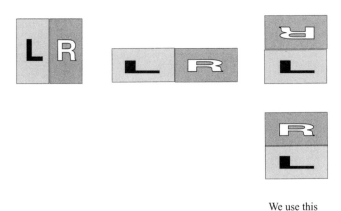

We use this

Figure 3.2. Kneading 2-d dough

We note that, as shown in Figure 3.1, the right rectangle ends up upside-down. For the sake of mathematical convenience, instead of putting the top piece upside-down, we will put it rightside-up. (From the point of view of the dynamical ideas we discuss, this change gives an equivalent process.) We are ready to define the 2-dimensional baker's transformation in detail. Start with a unit square. Cut the square down the middle to obtain two equal pieces (subrectangles). Then squeeze and stretch the left piece and do the same with the right piece; finally put the right piece on top of the left to bring us back to a square. This concludes one iteration of the process (Figure 3.2) and defines a transformation T of the unit square. It is easy to see that if a point is in the left subrectangle, then its horizontal distance is doubled and its vertical distance is halved, and points in the right subrectangle undergo a similar transformation, so T is given by the following formula:

$$T(x,y) = \begin{cases} (2x, \frac{y}{2}), & \text{if } 0 \leq x < 1/2; \\ (2x-1, \frac{y+1}{2}), & \text{if } 1/2 \leq x \leq 1. \end{cases}$$

3.1. An Example: The Baker's Transformation

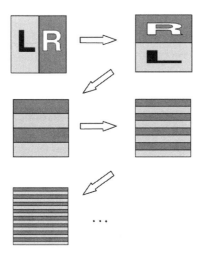

Figure 3.3. The baker's transformation

The first measurable dynamical property that T satisfies is that it preserves the area of subsets of the square. (Such a transformation is called *measure-preserving*.) Note that the rectangle L of area $1/2$ is sent to another rectangle $T(L)$ also of area $1/2$. The same is true for R and, furthermore, one can easily calculate that any rectangle whose sides are bounded by the *dyadic rationals* (i.e., numbers of the form $\frac{k}{2^n}$ for k and n integers) is sent to another rectangle of the same area. One can show that this property holds for arbitrary rectangles by approximating them with arbitrary dyadic rectangles. It is reasonable to ask what happens to more general subsets of the square. To this end, Chapter 2 develops the theory of Lebesgue measure, which generalizes the notion of area for the case of subsets of the plane. Our initial treatment of Lebesgue measure is restricted to subsets of the real line, but the same ideas generalize to subsets of $\mathbb{R}^d$. We show in Section 3.4 that to establish the measure-preserving property it suffices to verify it on a sufficiently large collection of sets, such as the collection of all rectangles with dyadic endpoints. We note here that in the definition of the measure-preserving property, as we shall see later, the condition is that the measure of the inverse image of a set is the same as the measure of the set; when the process or transformation is invertible, as is the case here, this is equivalent to the condition that the *forward* image of a set is the same as the measure of the set.

Studying what happens to specific sets when we iterate the baker's transformation helps understand the notion of mixing. Consider the set L (Figure 3.3). It is clear that $T(L)$ intersects L and R each in a square of area $1/4$; in other words, if we let $\lambda(X)$ denote the area (or measure) of a set X of the plane, we have that

$$\lambda(T(L) \cap A) = \lambda(L)\lambda(A),$$

where A is L or R. The reader is asked in the exercises to verify that for any sets A and B that are dyadic rectangles,

$$\lambda(T^n(A) \cap B) = \lambda(A)\lambda(B),$$

for all large enough integers $n \geq 0$. For more general sets, namely for any measurable sets A and B, it can be shown by approximating them by unions of dyadic rectangles and using the techniques of Chapter 6 that

(3.1) $$\lim_{n \to \infty} \lambda(T^n(A) \cap B) = \lambda(A)\lambda(B),$$

which, when $\lambda(B) \neq 0$, can be written as

(3.2) $$\lim_{n \to \infty} \frac{\lambda(T^n(A) \cap B)}{\lambda(B)} = \lambda(A).$$

This is the definition of *mixing* for a measure-preserving transformation. We interpret equation (3.2) as saying that the relative proportion in which the iterates of a given set A intersect a region of space B approximates the measure of A (Figure 3.4). For example, if in the kneading dough process one were to drop in some raisins occupying 5% of the area, and if we fix our gaze on a particular part of the space, then after some iterations one would expect to see approximately 5% of this region occupied by raisins.

Another important concept arises when we relax the notion of convergence in equation (3.1). It may be that convergence in the sense of (3.1) does not occur, but one may still have convergence in

3.1. An Example: The Baker's Transformation

the average, namely it may happen that for all measurable sets A and B it is the case that

$$(3.3) \qquad \lim_{n \to \infty} \frac{1}{n} \sum_{i=0}^{n-1} \lambda(T^i(A) \cap B) = \lambda(A)\lambda(B).$$

This is a weaker notion, and if (3.3) holds (for all measurable sets) we say that the transformation T is *ergodic*. We shall see that ergodicity is a weaker notion than mixing. It is also interesting to note, though we do not cover this, that by putting a topology on the set of measure-preserving transformations defined on a space such as the unit square, it can be shown that generically (i.e., for "a large set" in the sense of the topology) one will pick a transformation that is ergodic but not mixing. Another characterization of ergodicity can be seen by studying invariant sets: if there were to exist an invariant set A such that both A and its complement A^c had positive measure, then T would not be ergodic, because then $\lambda(T^n(A) \cap A^c) = 0$, contradicting (3.3) when $B = A^c$. The converse of this is also true but nontrivial and is a consequence of the ergodic theorem, which we prove in Chapter 5.

Returning to the modified 3-dimensional baker's transformation, as we saw that it admits an invariant set of positive measure whose complement is also of positive measure, it follows that the transformation is not ergodic. This example also shows an important property of such transformations. We have seen that the cube on which the modified 3-dimensional transformation is defined can be decomposed into cross sections of the form $x = c$ that are invariant, and the process is ergodic (in fact, mixing) with respect to 2-dimensional measure when restricted to each 2-dimensional slice. This is called the ergodic decomposition of the transformation and it is a special case of an important theorem called the ergodic decomposition theorem. In general terms, this theorem states that any measure-preserving transformation has a decomposition into ergodic components. Because of this theorem, when proving facts about transformations it is often possible to simply prove the result for the case of ergodic transformations.

This is a good place to mention a very useful property that lies properly between ergodicity and mixing. It may happen that the limit in (3.1) holds along a subsequence, that is, for each pair of measurable

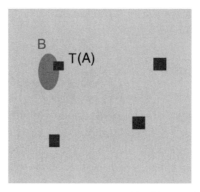

Figure 3.4. Mixing

sets A and B there may exist an increasing sequence n_i so that

(3.4) $$\lim_{i \to \infty} \lambda(T^{n_i}(A) \cap B) = \lambda(A)\lambda(B).$$

In this case we shall say that the transformation T is *weakly mixing*. Evidently, mixing implies weak mixing, and it is not hard to see that weak mixing implies ergodicity. In Chapter 6 we show that there exists a sequence n_i that "works" for all pairs of measurable sets A and B. Surprisingly, one can choose the sequence n_i to be of density 1 in the integers. Examples of transformations that are weakly mixing but not mixing are not immediately obvious and we shall construct some. Ergodic transformations that are not weakly mixing are easier to come by.

Exercises

(1) Show that in the case of the 2-dimensional baker's transformation T, for each dyadic rectangle A in the unit square, A and $T(A)$ have the same area.

(2) Verify that the mixing condition implies the ergodic condition.

3.2. Rotation Transformations

(3) State and prove the analogue of Exercise 1 for the case of the 3-dimensional baker's transformation.

3.2. Rotation Transformations

This section introduces one of the simplest examples of a transformation that leaves the Lebesgue measure of the unit interval invariant; such a transformation is said to be a *measure-preserving* transformation. A **transformation** is a function for which the domain and range are the same; in this case the image of a point is in the domain of the transformation. The simplest transformation on any set X is the identity transformation $\mathcal{I}$ defined by $\mathcal{I}(x) = x$ for all $x \in X$. If $T : X \to X$ is a transformation on a set X, as $T(x) \in X$ for all $x \in X$, the nth iterate of x, denoted $T^n(x)$, is defined by

$$T^0 = \mathcal{I},$$
$$T^{n+1} = T \circ T^n \text{ for } n \geq 0.$$

An **invertible transformation** is a transformation that is one-to-one and onto. In this case T^{-1} is also a transformation.

The rotation transformations we consider in this section are defined on the half-open unit interval $[0, 1)$. First we assign to each x in $\mathbb{R}$ a unique number in $[0, 1)$, denoted $(x \mod 1)$ and defined by

$$(x \mod 1) = x - \lfloor x \rfloor,$$

where $\lfloor x \rfloor$ is the largest integer $\leq x$. (So, $(x \mod 1) = x$ if $x \in [0, 1)$ and, for example, $(\pi \mod 1) = 0.1415....$) Using this one can define an equivalence relation on $\mathbb{R}$: two numbers $x, y \in \mathbb{R}$ are said to be **equivalent mod** 1 and written $x \equiv y \pmod{1}$ if $(x - y \mod 1) = 0$.

For any number $\alpha \in \mathbb{R}$, define the **rotation by** α to be the transformation

$$R_\alpha : [0, 1) \to [0, 1)$$

given by

$$R_\alpha(x) = (x + \alpha \mod 1).$$

When α is fixed and evident from the context we shall write R for R_α. Clearly, R_α is an invertible transformation.

Question. Show that for any $\alpha \in \mathbb{R}$ there exists $\alpha' \in [0,1)$ such that $R_\alpha = R_{\alpha'}$.

From now on we assume that $\alpha \in [0,1)$.

Example. If $\alpha = \frac{3}{4}$, then $R_\alpha(0) = \frac{3}{4}, R_\alpha^2(0) = R_\alpha(\frac{3}{4}) = (\frac{3}{4} + \frac{3}{4} \mod 1) = \frac{1}{2}$, $R_\alpha^3(0) = (\frac{1}{2} + \frac{3}{4} \mod 1) = \frac{1}{4}, R_\alpha^4(0) = R_\alpha(\frac{1}{4}) = 0$.

This example motivates one of the first notions in dynamics. If $T : X \to X$ is a transformation, then we call the set $\{T^n(x)\}_{n \geq 0}$ the **positive orbit** of x. When T is invertible we usually consider the **full orbit** of a point x, namely the set $\{T^n(x)\}_{n=-\infty}^{\infty}$.

Question. Show that if α is a rational number, then the orbit of every point x in $[0,1)$ under R_α consists of a finite number of points.

We say that a point $x \in X$ is a **periodic point** under a transformation $T : X \to X$ if $T^n(x) = x$ for some integer $n > 0$. The integer n is called a **period** of x; the **least period** is the smallest such integer n. A transformation is said to be a **periodic** transformation if every point is a periodic point for T; it is **strictly periodic** if every point has the same period. For example, $R_{3/4}$ is a strictly periodic transformation of least period 4.

Figure 3.5 shows the graph of R_α. The figure also suggests a description of R_α as an "interval exchange." Note that there are two partitions of $[0,1)$, one consisting of the subintervals $[0, 1-\alpha)$ and $[1-\alpha, 1)$, and the other of the subintervals $[0, \alpha)$ and $[\alpha, 1)$. The transformation R_α sends an interval in the first partition to an interval of the same length in the second partition.

Observe that

$$R_\alpha(0) = \alpha, \; R_\alpha(1-\alpha) = 0.$$

Also, when $x < 1 - \alpha$, then $x + \alpha < 1$, so $R_\alpha(x) = (x + \alpha \mod 1) = x + \alpha$. Therefore, R_α on $[0, 1-\alpha)$ is just the translation that sends x to $x + \alpha$ and whose graph is the first straight line of Figure 3.5. Now if $1 - \alpha < x < 1$, then $(x + \alpha \mod 1) = x + \alpha - 1$ and the graph of R_α is the second line of Figure 3.5. (Note that in the graph all

3.2. Rotation Transformations

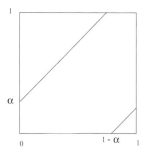

Figure 3.5. Rotation by R_α

intervals are open at 1.) From this it is clear that R_α is an invertible transformation on $([0, 1), \mathfrak{L}, \lambda)$.

Given a transformation $T : X \to X$ and a set $A \subset X$, we shall be interested in the set of points x such that $T(x) \in A$. Recall that the **inverse image** (or **pre-image**) of A is defined to be the set

$$T^{-1}(A) = \{x : T(x) \in A\}.$$

So, $T(x) \in A$ if and only if $x \in T^{-1}(A)$. In addition to considering the iterates of points $x, T(x), T^2(x), \ldots$ we consider the pre-images $T^{-n}(A), n \geq 0$. Note that $T^n(x) \in A$ if and only if $x \in T^{-n}(A)$. When T is invertible we shall also consider $T^n(A)$ for $n \geq 0$.

We now introduce the following definitions, which will be discussed in more detail in Section 3.4.

If $(X, \mathcal{S}, \mu)$ is a measure space, a transformation $T : X \to X$ is said to be a **measurable transformation** if the set $T^{-1}(A)$ is in $\mathcal{S}(X)$ for all $A \in \mathcal{S}(X)$. An **invertible measurable transformation** is an invertible transformation T such that T and T^{-1} are measurable. A transformation T is **measure-preserving** if it is measurable and

$$\mu(T^{-1}(A)) = \mu(A)$$

for all sets $A \in \mathcal{S}(X)$. In this case we say that μ is an **invariant measure** for T. Thus, a measure-preserving transformation must be measurable, but sometimes for emphasis we may describe a transformation as being measurable and measure-preserving. (An **invertible measure-preserving** transformation is an invertible measurable transformation that is measure-preserving.)

Question. Show that for any measure space $(X, \mathcal{S}, \mu)$ the identity transformation $\mathfrak{I} : X \to X$, defined by $\mathfrak{I}(x) = x$, is an invertible measure-preserving transformation.

Question. Let T be an invertible measurable transformation. Show that T is measure-preserving if and only if T^{-1} is measure-preserving.

Theorem 3.2.1. *The transformation R_α is an invertible measure-preserving transformation of $([0,1), \mathfrak{L}, \lambda)$.*

Proof. Recall that we may assume $\alpha \in [0,1)$. Note that the inverse of R_α is $R_\alpha^{-1} = R_{-\alpha}$, another rotation. Put $A_0 = A \cap [0, \alpha)$ and $A_1 = A \cap [\alpha, 1)$. Then note that $R_{-\alpha}(A_0)$ is precisely the translation by $-\alpha$ of the set A_0, which by Exercise 2.3.1 is measurable and of the same measure as A_0. A similar argument applies to A_1 and, since $A = A_0 \sqcup A_1$, this completes the proof. $\square$

As we shall see later, the proof of the measure-preserving property for a given transformation will usually be obtained as a consequence of Theorem 3.4.1, by which it suffices to show that $R_\alpha^{-1}(A)$ is measurable and that $\lambda(R_\alpha^{-1}(A)) = \lambda(A)$ for all measurable sets A in some sufficient semi-ring. For example, in the case of Theorem 3.2.1, the sufficient semi-ring can be the collection of intervals contained in $[0,1)$.

Our next theorem, Kronecker's theorem for irrational rotations, has several applications. Before we discuss its proof, we need to introduce some notation. We represent rotations as being on the unit interval, rather than on the unit circle, but we identify 0 with 1. Under this identification, the number 0.1 is closer to 0.9 than to 0.4. We now define a metric, or distance, that reflects the fact that we are considering numbers modulo 1. For $x, y \in [0, 1)$ define $d(x, y)$ by

$$d(x,y) = \min\{|x-y|, 1 - |x-y|\}.$$

Note that if $|x - y| \leq 1/2$, then $d(x, y) = |x - y|$.

3.2. Rotation Transformations

The proof of the following proposition is left to the reader.

Proposition 3.2.2. *The function d is a metric on $[0,1)$, and is such that if a sequence converges in the Euclidean metric, then it converges in the d metric. Furthermore, d is invariant for rotations, i.e., for any rotation R_α and any $x, y \in [0,1)$, $d(R_\alpha(x), R_\alpha(y)) = d(x,y)$.*

Theorem 3.2.3 (Kronecker). *If α is irrational, then for every $x \in [0,1)$ the sequence $\{R_\alpha^n(x)\}_{n \geq 0}$ is dense in $[0,1)$.*

Proof. We first show that when α is irrational, all the points $R_\alpha^n(x)$ are distinct for different integers n. Let R denote R_α. Now if $R^n(x) = R^m(x)$, for some integers m, n, then $x + n\alpha \equiv x + m\alpha \pmod 1$, so $(n-m)\alpha \equiv 0 \pmod 1$, which implies that $(n-m)\alpha$ is an integer. As α is irrational, this happens only when $m = n$. Therefore all the points in the orbit of x are distinct.

By the Bolzano–Weierstrass theorem we know that the sequence $\{R^n(x)\}_{n \geq 0}$ has a convergent subsequence in $[0,1]$. This implies that there are points in the orbit that are arbitrarily close to each other: given any $1/2 > \varepsilon > 0$ there exist nonnegative integers $p > q$ such that $0 < |R^p(x) - R^q(x)| < \varepsilon$. As $\varepsilon < 1/2$, $d(R^p(x), R^q(x)) < \varepsilon$. By Proposition 3.2.2, $d(R^k(y_1), R^k(y_2)) = d(y_1, y_2)$ for all $y_1, y_2 \in [0,1)$ and all integers k. So, $d(R^{p-q}(x), x) < \varepsilon$. Let $r = p - q$ and $\delta = d(R^{p-q}(x), x) > 0$. We claim that consecutive terms in the orbit $\{R^{\ell r}(x)\}_{\ell \geq 0}$ are δ-apart of each other. In fact, for $\ell \geq 0$,

$$d(R^{(\ell+1)r}(x), R^{\ell r}(x)) = d(R^r(x), x) = \delta < \varepsilon.$$

This shows that the distinct points $\{R^{\ell r}(x)\}_{\ell=0}^\infty$ subdivide $[0,1)$ into subintervals of length $< \varepsilon$. Therefore, $\{R^n(x)\}_{n \geq 0}$ is dense in $[0,1)$. □

The notion of continuity of real functions, with which we assume the reader is familiar, generalizes in a natural way to metric spaces. Let (X, d) and (Y, q) be two metric spaces. A map $\phi : X \to Y$ is said to be **continuous at a point** $x \in X$ if for all $\varepsilon > 0$ there exists a number $\delta > 0$ such that $q(\phi(x), \phi(y)) < \varepsilon$ whenever $y \in X$ and $d(x, y) < \delta$. We say that f is **continuous** on X if f is continuous at x for every point x in X. The reader is asked to show that R_α is continuous with respect to the d metric of Propostion 3.2.2 (Exercise 6).

Theorem 3.2.3 illustrates an important concept. If (X, d) is a metric space, a map $T : X \to X$, usually continuous, is defined to be **minimal** if the positive orbit $\{T^n(x)\}_{n \geq 0}$ is dense in X for all $x \in X$. Kronecker's theorem states that irrational rotations are minimal. Minimality is not invariant under (measure-preserving) isomorphisms (defined in Section 3.10).

Application 1. We apply Theorem 3.2.3 to an interesting question known as Gelfand's question. This question is concerned with the first digits of powers of 2. Here is a list of the first 20 powers of 2:

$$2, 4, 8, 16, 32, 64, 128, 256, 512, 1024, 2048, 4096, 8192,$$
$$16384, 32768, 65536, 131072, 262144, 524288, 1048576$$

The sequence of first digits of the first 40 powers of 2 is:

$$2, 4, 8, 1, 3, 6, 1, 2, 5, 1,$$
$$2, 4, 8, 1, 3, 6, 1, 2, 5, 1,$$
$$2, 4, 8, 1, 3, 6, 1, 2, 5, 1,$$
$$2, 4, 8, 1, 3, 6, 1, 2, 5, 1.$$

Do we ever see a 7, a 9? Gelfand's question asks: how often do we see a power of 2 that starts with a 7, and with what frequency? We show here that there are infinitely many integers n such that 2^n starts with a 7. Surprisingly, they have a well-defined frequency. The existence of this frequency follows from the uniform distribution of multiples of an irrational number modulo 1. While this fact can be given an independent proof, we will obtain it as a consequence of the ergodic theorem, which we study in Chapter 5. We shall see later in Section 5.3, that the digit $d, 0 < d < 10$, occurs as a first digit in powers of 2 with frequency $\log_{10}(d+1)/d$. So for example the digit 1 appears with frequency about 0.3. Benford's Law is a statement asserting that the digit 1 occurs as a first digit in many naturally occurring tables, such as street addresses, stock prices, electric bills, with frequency about $1/3$ (rather than about $1/9$ as might be expected), and that the frequency of the other digits continues to be distributed in this logarithmic form. This was first observed by S. Newcomb in 1881

3.2. Rotation Transformations

regarding first digits in logarithm tables, but only justified recently in 1996 by T. Hill.

We start with the following simple observation. As the integer 721, say, starts with a 7 there is some integer $k > 0$ so that

$$7 \times 10^k \leq 721 < 8 \times 10^k.$$

To generalize this, let $d \in \{1, \ldots, 9\}$ (a similar analysis can be done for any integer $d > 0$, but the details are left to the reader). If the decimal representation of the integer 2^n starts with d, then for some integer $k \geq 0$,

$$d \times 10^k \leq 2^n < (d+1) \times 10^k.$$

Thus

$$\log_{10}(d \times 10^k) \leq \log_{10} 2^n < \log_{10}((d+1) \times 10^k).$$

In other words,

$$\log_{10} d \leq n \log_{10} 2 - k < \log_{10}(d+1), \text{ or}$$
$$\log_{10} d \leq n \log_{10} 2 \bmod 1 < \log_{10}(d+1).$$

But this is the same as saying that, letting $\alpha = \log_{10} 2$,

$$R_\alpha^n(0) \in [\log_{10} d, \log_{10}(d+1)).$$

Since $\alpha = \log_{10} 2$ is irrational, by Theorem 3.2.3 there are infinitely many integers n such that $R_\alpha^n(0) \in [\log_{10} d, \log_{10}(d+1))$. Thus there are infinitely many powers of 2 that start with a 7.

Application 2. (A Nonmeasurable Set.) The following construction of a nonmeasurable set is based on the Axiom of Choice. We first recall the statement of this axiom: Given any collection of nonempty sets A_α indexed by some nonempty set Γ, there exists a function F, called a choice function, whose domain is Γ and whose range is $\bigcup_{\alpha \in \Gamma} A_\alpha$, such that $F(\alpha) \in A_\alpha$, for each $\alpha \in \Gamma$. We think of F as choosing an element of A_α for each $\alpha \in \Gamma$. If the set Γ were a finite set, using mathematical induction on the integers, it is not hard to give a proof for this axiom. However, P.J. Cohen showed in 1963 that the Axiom of Choice is independent of the standard axioms of set theory (the Zermelo-Fraenkel axioms), i.e., neither it nor its negation can be deduced from these axioms. Furthermore, in 1970 Solovay showed that it cannot be shown that there exist nonmeasurable sets of reals with the Zermelo-Fraenkel axioms of set theory without the Axiom of Choice, on the assumption that the existence of *inaccessible cardinals* is consistent with the Zermelo-Fraenkel axioms. The Axiom of Choice is a reasonable axiom that is assumed by most mathematicians.

We use irrational rotations to show the existence of a non-measurable set. Let R be any rotation by an irrational number. Then for any $x \in [0, 1)$ the full orbit $\Gamma_x = \{R^n(x)\}_{n=-\infty}^{\infty}$ consists of distinct points. We claim that the collection of orbits forms a partition of $[0, 1)$, i.e., we claim that every point in $[0, 1)$ is in some orbit and that if any two orbits intersect in a nonempty set, then they must be equal. Indeed, for any $x \in [0, 1)$, $x \in \Gamma_x$, and if $z \in \Gamma_x \cap \Gamma_y$, for some $x, y \in [0, 1)$, then $z = R^n(x)$ and $z = R^m(y)$ for some integers m, n. So $x = R^{m-n}(y)$ which means that $\Gamma_x = \Gamma_y$. (Another way to obtain this partition is to define an equivalence relation on $[0, 1)$ by declaring that two points x, y are equivalent if $x \in \Gamma_y$. One verifies that this is an equivalence relation and that the equivalence classes are the orbits.)

As each orbit is nonempty, we can use the Axiom of Choice to construct a set E consisting of precisely one point from each orbit. It follows that for each integer n, $R^n(E)$ also consists of one point from each orbit, and this means that the collection $\{R^n(E)\}_{n=-\infty}^{\infty}$ forms a countable partition of $[0, 1)$ into disjoint sets.

We now show that assuming that the set E is measurable leads to a contradiction. So suppose that E is measurable. Since R is measure-preserving, $R^n(E)$ is measurable and $\lambda(R^n(E)) = \lambda(E)$ for all n. Since $\bigcup_{n=-\infty}^{\infty} R^n(E) = [0, 1)$, by Countable Additivity

$$1 = \sum_{n=-\infty}^{\infty} \lambda(R^n(E)) = \sum_{n=-\infty}^{\infty} \lambda(E).$$

But if $\lambda(E) = 0$, then $\sum_{n=-\infty}^{\infty} \lambda(E) = 0$ and if $\lambda(E) > 0$, then $\sum_{n=-\infty}^{\infty} \lambda(E) = \infty$. So in both cases we reach a contradiction. Therefore the set E is not measurable.

Exercises

(1) Let d be any positive integer. Show that there are infinitely many positive integers n so that 2^n starts with d.

3.3. The Doubling Map

(2) Show that there are infinitely many positive integers n so that 3^n starts with 1984.

(3) Let $X_n = \{x_0, \ldots, x_{n-1}\}$ and let μ be the counting measure on X_n, i.e., $\mu(\{x_i\}) = 1$ for $i = 0, \ldots, n-1$. Define a transformation T on X_n by $T(x_i) = x_{i+1}$ if $i = 0, \ldots, n-2$, and $T(x_{n-1}) = x_0$. Show that T is a measure-preserving invertible transformation on (X_n, μ). T is called a rotation on n points.

(4) Give another proof of Kronecker's Theorem by showing that if the orbit of a point $x \in [0, 1)$ is not dense, then one can choose an open interval in the complement of the orbit and then reach a contradiction.

(5) Prove Proposition 3.2.2.

(6) Show that R_α is continuous with respect to the metric d.

(7) Show that, assuming the Axiom of Choice, every interval of positive length contains a nonmeasurable set. (In Exercise 3.11.1 you will show that every set of positive measure contains a non-measurable set.)

(8) Is it the case that for every $\varepsilon > 0$ there exists a non-Lebesgue measurable set E with $\lambda^*(E) < \varepsilon$?

* (9) A subset E of $\mathbb{R}$ is called a **Bernstein set** if both E and E^c have a nonempty intersection with each uncountable closed set. Show that a Bernstein set, if it exists, is non-Lebesgue measurable. (For a construction of Bernstein sets see [**56**].)

* (10) Are there infinitely many positive integers n so that both 2^n and 3^n start with 7?

3.3. The Doubling Map: A Bernoulli Noninvertible Transformation

The second example that we study is another transformation defined on $[0, 1)$; however this time it is not invertible, but two-to-one. Define the transformation T on $[0, 1)$ by

$$T(x) = 2x \ (\text{mod } 1) = \begin{cases} 2x, & \text{if } 0 \leq x \leq 1/2; \\ 2x - 1, & \text{if } 1/2 < x < 1. \end{cases}$$

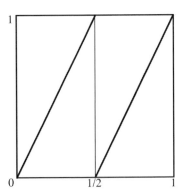

Figure 3.6. The doubling map

One can think of T as a one-dimensional version of the baker's transformation. We call T the **doubling map** transformation. Before studying its properties we investigate a new representation of the doubling map.

Let D consist of all the numbers in $[0,1]$ of the form $\frac{k}{2^n}$, and let $X_0 = [0,1] \setminus D$. The numbers in $[0,1] \setminus D$ have a unique representation in binary form as

$$x = \sum_{i=1}^{\infty} \frac{a_i}{2^i}.$$

We call the sequence $(a_1 a_2 \cdots)$ the **symbolic binary representation** of x. Then the doubling map for points $x \in [0,1] \setminus D$ is given by

$$T(x) = \sum_{i=1}^{\infty} \frac{a_{i+1}}{2^i}.$$

That is, T is a *shift* of the representation of x, i.e., the point in $[0,1)$ whose symbolic binary representation is $(a_1 a_2 a_3 \ldots)$ is sent to the point with symbolic binary representation $(a_2 a_3 \ldots)$. The following theorem shows that the doubling map is measure-preserving. The use of the inverse image of sets in the definition of the measure-preserving property is further clarified in Lemma 4.4.6.

Theorem 3.3.1. *The doubling map transformation T is a measure-preserving transformation on $([0,1), \mathfrak{L}, \lambda)$. Furthermore, the set of periodic points of T is dense in $[0,1)$.*

3.3. The Doubling Map

Proof. We first show it is measure-preserving. Define the maps $S_1 : [0,1) \to [0, \frac{1}{2})$ and $S_2 : [0,1) \to [\frac{1}{2}, 1)$ by $S_1(y) = y/2$ and $S_2(y) = y/2 + 1/2$. We saw in Exercise 2.3.1 that for any measurable set A, the sets $\frac{1}{2}A$ and $\frac{1}{2}A + \frac{1}{2}$ are measurable and $\lambda(\frac{1}{2}A) = \frac{1}{2}\lambda(A) = \lambda(\frac{1}{2}A + \frac{1}{2})$. Next we observe that $T^{-1}(A) = S_1(A) \sqcup S_2(A)$. So $\lambda(T^{-1}(A)) = \lambda(S_1(A)) + \lambda(S_2(A)) = \lambda(\frac{1}{2}A) + \lambda(\frac{1}{2}A + \frac{1}{2}) = \frac{1}{2}\lambda(A) + \frac{1}{2}\lambda(A) = \lambda(A)$.

A second proof can be given using Theorem 3.4.1.

For the periodic points, one can verify that for each integer $p \geq 1$ the points of period p are those whose symbolic binary representation consists of the infinitely repeated string $a_1 a_2 \ldots a_p$ for any $a_1, \ldots, a_p \in \{0, 1\}$. The set of these points is clearly dense in $[0, 1)$. $\square$

Application. From the representation of T as a shift of the binary numbers in $[0, 1)$, it follows that

$$T^i(x) \in [0, 1/2)$$

if and only if the ith digit in the binary expansion of x is 0, and

$$T^i(x) \in [1/2, 1)$$

if and only if the ith digit in the binary expansion of x is 1.

Using the important **characteristic function** or **indicator function** of a set A, denoted by $\mathbb{I}_A$ and defined by

$$\mathbb{I}_A(x) = \begin{cases} 1, & \text{if } x \in A; \\ 0, & \text{if } x \notin A, \end{cases}$$

we see that the ith digit in the binary expansion of x is 1 if and only if $T^i(x) \in [1/2, 1)$, and it is 0 if and only if $T^i(x) \in [0, 1/2)$.

Therefore, the frequency of appearances of 0 in the binary representation of x, if it exists, can be expressed as the limit

(3.5) $$\lim_{n \to \infty} \frac{1}{n} \sum_{i=0}^{n} \mathbb{I}_{[0, 1/2)}(T^i(x)).$$

Limits of the above form will be studied in Chapter 5, and their existence, for all x outside a set of measure zero, will be a consequence of the Ergodic Theorem. A number x is said to be a **simply normal number** to the base 2 if the limit in (3.5) exists and equals $1/2$. A consequence of the ergodic theorem will be that almost all (i.e., outside a null set) numbers are *normal*, an extension of simply normal defined in Section 5.2. Normal numbers are discussed in more detail in Section 5.2.

Note that the only point of discontinuity of T is at $x = 1/2$. However, with respect to the metric d of Section 3.2, T is continuous. We end with another example that has similar properties but is continuous with respect to the Euclidean metric. Define the **tent map** $T : [0,1] \to [0,1]$ by

$$T(x) = \begin{cases} 2x, & \text{if } x \in [0, 1/2); \\ 2 - 2x, & \text{if } x \in [1/2, 1]. \end{cases}$$

Let (X, d) be a metric space and let $T : X \to X$ be a map, usually a continuous map. The map T is **topologically transitive** (sometimes called one-sided topologically transitive) if there is a point $x \in X$ such that its positive orbit $\{T^n(x) : n \geq 0\}$ is dense in X. For example, the tent map T is topologically transitive (Exercise 8). Also, any minimal map is topologically transitive.

We conclude with a very remarkable theorem that has proven very useful in establishing existence results in dynamics. We first need a few definitions. The notion of nowhere dense sets that we have seen earlier can be generalized to metric spaces. Let (X, d) be a metric space. A set A in X is said to be **nowhere dense** if its closure contains no nonempty open sets; we saw, for example, that the Cantor set in $[0, 1]$ is nowhere dense. We observe that A is nowhere dense if and only if for any nonempty open set G there is a nonempty open set H contained in $G \setminus A$. In fact, suppose A is nowhere dense and G is a nonempty open set. If $G \setminus A$ were to contain no open sets, then A would be dense in G, or the closure of A would contain G, a contradiction. Conversely, if for any nonempty open set G there is a nonempty open set $H \subset G \setminus A$, then A cannot be dense in any open set G, which means that its closure contains no nonempty open sets.

A set is said to be **meager** or of **first category** if it is a countable union of nowhere dense sets; for example, the set of rational numbers in $\mathbb{R}$, while not nowhere dense is a meager set. The concept of meager captures the notion of being topologically small.

3.3. The Doubling Map

The next theorem basically states that a complete metric space cannot be topologically small. In fact, it says that a nonempty open set in a complete metric space cannot be meager.

Lemma 3.3.2. *Let (X, d) be a metric space. Then the following are equivalent.*

(1) *Every nonempty open set is not a meager set.*

(2) *The intersection of any countable collection of dense open sets is dense.*

Proof. Suppose that a nonempty open set is not meager. Let $A = \bigcap_{i=1}^{\infty} G_i$, where each G_i is an open dense set. Then G_i^c is nowhere dense. If A were not dense there would exist a nonempty open set H contained in A^c. But $A^c = \bigcup_{i=1}^{\infty} G_i^c$, a meager set. This contradicts that it contains H.

For the second part suppose that H is an open set that is meager. Then $H = \bigcup_{i=1}^{\infty} E_i$, where the E_i are nowhere dense sets. Then their closures $\bar{E}_i$ are also nowhere dense. Then $H \subset \bigcup_{i=1}^{\infty} \bar{E}_i$, so H^c contains $\bigcap_{i=1}^{\infty} \bar{E}_i^c$, a countable intersection of dense open sets, which cannot be dense as it does not intersect H, a contradiction. □

Theorem 3.3.3 (Baire Category Theorem). *Let X be a complete metric space. Then the intersection of any countable collection of dense open sets in X is dense.*

Proof. Let $\{G_n\}$ be a countable collection of dense open sets in X. We show that if B is any open ball of positive radius, then it intersects $\bigcap_{n=1}^{\infty} G_n$. The idea is to construct a point in the intersection as the limit of a Cauchy sequence.

We start the construction of the sequence. For this we have to be careful to choose the sequence inside balls of decreasing radius ε_n to guarantee it is Cauchy—for example, it suffices to take $\varepsilon_n < \frac{1}{n}$. First note that as G_1 is dense, B has a nonempty intersection with G_1. Choose a ball, of the form $B(x_1, \varepsilon_1)$, so that its closure is contained in $B \cap G_1$ and with $\varepsilon_1 < 1$. The ball $B(x_1, \varepsilon_1)$ must also intersect G_2 and we may similarly choose a ball $B(x_2, \varepsilon_2)$ so that its closure is contained in $B(x_1, \varepsilon_1) \cap G_2$ and with $\varepsilon_2 < 1/2$. In this way we can generate a sequence of balls $B(x_n, \varepsilon_n)$ so that $\overline{B}(x_n, \varepsilon_n) \subset B(x_{n-1}, \varepsilon_{n-1}) \cap G_n$

and $\varepsilon_n < 1/n$. We verify that the sequence $\{x_n\}$ is a Cauchy sequence: for any $\varepsilon > 0$ choose $m > 0$ so that $2\varepsilon_m < \varepsilon$. Then all the points x_k, for $k \geq m$, are in $B(x_m, \varepsilon_m)$. So for any $k, \ell \geq m$

$$d(x_k, x_\ell) \leq d(x_k, x_m) + d(x_m, x_\ell) < 2\varepsilon_m < \varepsilon.$$

Therefore the sequence converges to a point x in X. Clearly, $x \in \bigcap_{n=1}^\infty \overline{B}(x_n, \varepsilon_n)$. Finally we observe that by construction

$$\emptyset \neq \bigcap_{n=1}^\infty \overline{B}(x_n, \varepsilon_n)) \subset \bigcap_{n=1}^\infty G_n \cap B.$$

□

As an application we prove a characterization of transitive maps.

Theorem 3.3.4. *Let (X, d) be a complete, separable, metric space without isolated points. Let $T : X \to X$ be a continuous transformation. Then the following are equivalent.*

(1) *T is topologically transitive.*

(2) *The set of points in X that have a dense positive orbit is a dense $\mathcal{G}_\delta$ set.*

(3) *For all nonempty open sets U and V there exists an integer $n > 0$ such that $T^{-n}(U) \cap V \neq \emptyset$.*

(4) *For all nonempty open sets U and V there exists an integer $n > 0$ such that $T^n(U) \cap V \neq \emptyset$.*

Proof. (1) $\Rightarrow$ (4): Let x be a point with a dense positive orbit. Then there exists $k > 0$ so that $T^k(x) \in U$. Let $u = T^k(x)$. As X has no isolated points u has a dense positive orbit (Exercise 9). Then there exists $n > 0$ such that $T^n(u) \in V$. This means $T^n(U) \cap V \neq \emptyset$.

(4) $\Rightarrow$ (3): There exists $n > 0$ so that $T^n(U) \cap V \neq \emptyset$. This means that there is an element of $T^n(U)$, which must have the form $T^n(u)$ for some $u \in U$, that is in V. So u is in $T^{-n}(V)$ and also in U. Thus, $U \cap T^{-n}(V) \neq \emptyset$.

(3) $\Rightarrow$ (2): As X is separable, we can choose a sequence $\{x_m\}$ that is dense in X. Let $\{r_k\}$ be a countable sequence decreasing to 0

3.3. The Doubling Map

(say, $r_k = 1/k$). Note that if
$$x \in \bigcap_{m,k \geq 1} \bigcup_{n \geq 1} T^{-n}(B(x_m, r_k)),$$
then for all $m, k \geq 1$, $T^n(x) \in B(x_m, r_k)$ for some $n \geq 1$. This means that the positive orbit of x is dense. By the hypothesis, for all $m, k \geq 1$, any nonempty open set U must intersect $\bigcup_{n \geq 1} T^{-n}(B(x_m, r_k))$, so it follows that the set $\bigcup_{n \geq 1} T^{-n}(B(x_m, r_k))$ is dense. As the set $\bigcap_{m,k \geq 1} \bigcup_{n \geq 1} T^{-n}(B(x_m, r_k))$ is a countable intersection of dense open sets, the Baire category theorem implies that this set is dense. So the set of points with a dense orbit is dense.

Finally, it is clear that (2) $\Rightarrow$ (1). □

Regarding completeness for metric spaces we observe that there can be two metrics that generate the same open sets for a space X but one is complete and the other is not.

Exercises

(1) Let $T : [0,1] \to [0,1]$ be defined by $T(x) = 2x$ if $0 \leq x \leq 1/2$ and $T(x) = 2 - 2x$ if $1/2 < x \leq 1$. Show that T is finite measure-preserving. Find a point $x \in [0,1)$ such that the (positive) orbit of x under T is dense.

(2) For an integer $k > 1$ define $T_k(x) = kx \pmod{1}$ for $x \in [0,1)$. Show that T_k is measure-preserving for Lebesgue measure. Find other measures in $[0,1)$ that are invariant under T_k.

(3) Let T be the doubling map. Show that the set of points that are periodic for T is a dense set in $[0,1)$.

(4) Let T be the doubling map. Find a point $x \in [0,1)$ whose T-orbit is dense.

(5) Show that the doubling map is continuous with respect to the metric d of Section 3.2.

(6) Show that the doubling map is topologically transitive.

(7) Let $f(x) = 4x(1-x)$ be defined in $[0,1]$. Is f topologically transitive? Show that it has infinitely many periodic points.

(8) Show that the tent map T is topologically transitive.

(9) Let $T : X \to X$ be a continuous map of a metric space with no isolated points and let $p \in X$. Show that if x is in the closure of the positive orbit of p, then x is an accumulation point of the positive orbit of p. Conclude that if p has a positive dense orbit, then every point in the positive orbit of p has a positive dense orbit.

(10) Let X be a complete separable metric space with no isolated points and let $T : X \to X$ be a homeomorphism. Show that if there is a point $x \in X$ such that its orbit $\{T^n(x)\}_{n=-\infty}^{\infty}$ is dense, then T is (one-sided) topologically transitive.

(11) Let (X, d) be a complete metric space. Show that T is topologically transitive if and only if for any closed set F such that $F \subset T^{-1}(F)$, then $F = X$ or F is nowhere dense.

(12) Let A be a $\mathcal{G}_\delta$ subset of a metric space. Show that if A is dense, then its complement A^c is meager.

(13) A set A in $\mathbb{R}$ is said to be **residual** if its complement is a meager set. Show that the set of Liouville numbers is residual. Deduce that $\mathbb{R}$ can be written as the disjoint union of a null set and a meager set.

(14) Let X be a nonempty set. A σ-**ideal** on X is a collection of subsets of X that contains $\emptyset$ and is closed under subsets and countable unions. Let (X, d) be a metric space. Show that the collection of meager sets in X is a σ-ideal.

(15) Construct an open dense subset of $[0, 1]$ of arbitrarily small measure. (Hint: Let $\varepsilon > 0$ and let q_n be a countable dense subset in $[0, 1]$. Choose an open ball of radius $\varepsilon/2^n$ around each $q_n, n \geq 1$.)

(16) Find all periodic points for the tent map.

(17) Show that there cannot exist an invertible continuous map of the interval $[0, 1]$ that is topologically transitive.

3.4. Measure-Preserving Transformations

This section studies measure-preserving transformations in more detail. Let $(X, \mathcal{S}, \mu)$ be a measure space. We shall call a measure-preserving transformation $T : X \to X$ **finite measure-preserving** if $\mu(X) < \infty$ and **infinite measure-preserving** otherwise. When $\mu(X) = 1$, we may say that T is a **probability-preserving** transformation. In the infinite measure-preserving case we shall only consider the case when X is σ-finite. We say that $(X, \mathcal{S}, \mu, T)$ is a **measure-preserving dynamical system** if $(X, \mathcal{S}, \mu)$ is a σ-finite measure space and $T : X \to X$ is a measure-preserving transformation.

The main result is that to show that a transformation is measure-preserving it suffices to check the measure-preserving property on a sufficient semi-ring.

In applications, sufficient semi-rings will often be some collection of intervals such as the (left-closed, right-open) dyadic intervals. While the statement of the theorem is important, the proof may be omitted on a first reading.

Theorem 3.4.1. *Let $(X, \mathcal{S}, \mu)$ be a complete σ-finite measure space with a sufficient semi-ring $\mathcal{C}$. If for all I in $\mathcal{C}$,*

(1) $T^{-1}(I)$ *is a measurable set, and*

(2) $\mu(T^{-1}(I)) = \mu(I)$,

then T is a measure-preserving transformation.

Proof. It suffices to show that for any $A \in \mathcal{S}(X)$ with $\mu(A) < \infty$, $T^{-1}(A)$ is measurable and $\mu(T^{-1}(A)) = \mu(A)$.

The proof consists of two parts. In the first part we show that for each measurable set A, if $H(A)$ is as defined in Lemma 2.7.2, then $T^{-1}(H(A))$ is measurable and $\mu(T^{-1}(H(A))) = \mu(H(A))$. We know that

$$H(A) = \bigcap_{n=1}^{\infty} H_n,$$

with $H_n \supset H_{n+1}$ and $\mu(H_n) < \infty$. Furthermore, each H_n can be written as
$$H_n = \bigsqcup_{i=1}^{\infty} C_{n,i},$$
with $C_{n,i} \in \mathcal{C}$. Thus, the set
$$T^{-1}(H_n) = T^{-1}(\bigsqcup_{i \geq 1} C_{n,i}) = \bigsqcup_{i \geq 1} T^{-1}(C_{n,i})$$
is measurable. Also,
$$\mu(T^{-1}(H_n)) = \mu(\bigsqcup_{i \geq 1} T^{-1}(C_{n,i})) = \sum_{i \geq 1} \mu(T^{-1}(C_{n,i}))$$
$$= \sum_{i \geq 1} \mu(C_{n,i}) = \mu(\bigsqcup_{n \geq 1} C_{n,i}) = \mu(H_n).$$

Therefore $T^{-1}(H(A)) = T^{-1}(\bigcap_{n \geq 1} H_n) = \bigcap_{n \geq 1} T^{-1}(H_n)$ is measurable, and as $\mu(H_n) < \infty$, using Proposition 2.5.2,
$$\mu(T^{-1}(H(A))) = \mu(\bigcap_{n \geq 1} T^{-1}(H_n)) = \lim_{n \to \infty} \mu(T^{-1}(H_n))$$
$$= \lim_{n \to \infty} \mu(H_n) = \mu(\bigcap_{n \geq 1} H_n) = \mu(H(A)).$$

This concludes the proof of the first part.

For the second part of the proof we use that $A \subset H(A)$ and $\mu(H(A) \setminus A) = 0$. Write $N = H(A) \setminus A$.

Apply Lemma 2.7.2 again, this time to the null set N to obtain a null set $H(N)$ containing N. By the first part, $T^{-1}(H(N))$ is measurable and $\mu(T^{-1}(H(N))) = \mu(H(N)) = \mu(N) = 0$. As $N \subset H(N)$, then $T^{-1}(N) \subset T^{-1}(H(N))$, so $\mu(T^{-1}(N)) = 0$. Therefore $T^{-1}(N)$ is measurable.

Finally, note that
$$T^{-1}(A) = T^{-1}(H(A)) \setminus T^{-1}(N).$$

This shows that $T^{-1}(A)$ is measurable and, furthermore,
$$\mu(T^{-1}(A)) = \mu(T^{-1}(H(A))) - \mu(T^{-1}(N))$$
$$= \mu(H(A)) - 0 = \mu(A).$$

$\square$

3.4. Measure-Preserving Transformations

Second Proof: We give another shorter but nonconstructive proof of this theorem. Let
$$\mathcal{A} = \{A : A \in \mathcal{S} \text{ and } T^{-1}(A) \in \mathcal{S}, \mu(T^{-1}(A)) = \mu(A)\}.$$
Clearly $\mathcal{A}$ contains the semi-ring $\mathcal{C}$. One can verify that $\mathcal{A}$ is a monotone class. It follows that $\mathcal{A}$ contains $\mathcal{S}$, so T is measure-preserving, completing the proof.

Example. We give another proof that the Doubling Map is measure-preserving. Let $\mathcal{D}$ be the collection of left-closed right-open dyadic intervals in $[0,1)$. For I in $\mathcal{D}$, write $I = [k/2^i, (k+1)/2^i)$ for integers k, i with $i \geq 0$, $k \in \{0, \ldots, 2^i - 1\}$. Then
$$T^{-1}(I) = [\frac{k}{2^{i+1}}, \frac{k+1}{2^{i+1}}) \cup [\frac{k+2^i}{2^{i+1}}, \frac{k+1+2^i}{2^{i+1}}).$$
Evidently, $T^{-1}(I)$ is measurable and one can check that $\mu(T^{-1}(I)) = \frac{1}{2^i} = \mu(I)$. We saw in Section 2.7 that $\mathcal{D}$ is a sufficient semi-ring, so an application of Theorem 3.4.1 yields that T is measure-preserving.

Exercises

(1) (Boole's transformation) Let $T : \mathbb{R} \to \mathbb{R}$ be defined by $T(x) = x - \frac{1}{x}$ if $x \neq 0$ and $T(0) = 0$. Show that T is measure-preserving on $\mathbb{R}$ with Lebesgue measure.

(2) Show that if $(X, \mathcal{S}, \mu)$ is a σ-finite measure-space and $T : X \to X$ is measure-preserving, then for any $X_0 \in \mathcal{S}(X)$ with $T^{-1}(X_0) = X_0$, the system $(X_0, \mathcal{S}(X_0), \mu, T)$ is a measure-preserving dynamical system.

(3) Let $(X, \mathcal{S}, \mu)$ be a σ-finite measure space and let $X_0 \in \mathcal{S}(X)$ with $\mu(X \setminus X_0) = 0$. Suppose there exists a transformation T_0 so that $(X_0, \mathcal{S}(X_0), \mu, T_0)$ is a measure-preserving dynamical system. Show that there exists a transformation $T : X \to X$ so that $T(x) = T_0(x)$ for $x \in X_0$ and $(X, \mathcal{S}, \mu, T)$ is a measure-preserving dynamical system. (T is not unique but differs from T_0 on only a null set.)

(4) Show that if $(X, \mathcal{S}, \mu, T)$ is a measure-preserving dynamical system, then for any integer $n > 0$, $(X, \mathcal{S}, \mu, T^n)$ is a measure-preserving dynamical system.

(5) Show that if $(X, \mathcal{S}, \mu, T)$ is an invertible measure-preserving dynamical system, then for any integer n, $(X, \mathcal{S}, \mu, T^n)$ is an invertible measure-preserving dynamical system.

(6) Complete the details of the second proof of Theorem 3.4.1.

3.5. Recurrence

A measure-preserving transformation T defined on a measure space $(X, \mathcal{S}, \mu)$ is said to be **recurrent** if for every measurable set A of positive measure there is a null set $N \subset A$ such that for all $x \in A \setminus N$ there is an integer $n = n(x) > 0$ with

$$T^n(x) \in A.$$

Informally, T is recurrent when for any set A of positive measure almost every point of A returns to A at some future time. We think of n as a "return time" to A. Figure 3.7 shows a typical point in a set of positive measure for a recurrent transformation. We will see in Theorem 3.5.3 that every finite measure-preserving transformation is recurrent. However, the transformation $T: \mathbb{R} \to \mathbb{R}$ defined by $T(x) = x + 1$ is measure-preserving for Lebesgue measure, but is easily seen not to be recurrent (let $A = [0, 1)$, for example).

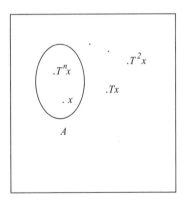

Figure 3.7. T is recurrent

As the following lemma shows, once a transformation is recurrent there exist infinitely many return times.

3.5. Recurrence

Lemma 3.5.1. *Let $(X, \mathcal{S}, \mu)$ be a measure space and $T : X \to X$ a recurrent measure-preserving transformation. Then, for all sets of positive measure A, there exists a null set N such that for all $x \in A \setminus N$ there is an increasing sequence $m_i > 0$ with $T^{m_i}(x) \in A \setminus N$ for all $i \geq 1$.*

Proof. By definition, there is a null set N_1 such that for all $x \in A \setminus N_1$ there is an integer $n = n(x)$ with $T^n(x) \in A$. Let $N = \bigcup_{k=0}^{\infty} T^{-k}(N_1)$. N is a null set as $\mu(T^{-k}(N_1)) = \mu(N) = 0$ for $k \geq 0$. For all $k \geq 0$, $T^{-k}(N) \subset N$, so if $x \notin N$, then $T^k(x) \notin N$. Thus, if we choose $x \in A \setminus N \subset A \setminus N_1$, then $T^{n(x)}(x) \in A \setminus N$. Let $m_1 = n(x)$. Applying the result we just proved to the point $z = T^{m_1}(x)$ that is in $A \setminus N$, we obtain an integer $n_1 = n(z) > 0$ so that $T^{n_1}(z) \in A \setminus N$. If we let $m_2 = m_1 + n_1$, then $T^{m_2}(x) = T^{n_1}(z) \in A \setminus N$. Continue in this manner to obtain an increasing sequence of integers m_i. □

The following lemma shows a useful characterization of recurrence. A measure-preserving transformation T on X is said to be **conservative** if for any set A of positive measure there exists an integer $n > 0$ such that $\mu(A \cap T^{-n}(A)) > 0$.

Lemma 3.5.2. *Let $(X, \mathcal{S}, \mu)$ be a measure space. A measure-preserving transformation T on X is recurrent if and only if it is conservative.*

Proof. First observe that T is recurrent if and only if

$$\mu(A \setminus \bigcup_{n=1}^{\infty} T^{-n}(A)) = 0$$

for all sets A of positive measure. This is so since the set $\bigcup_{k=1}^{\infty} T^{-k}(A)$ consists of all points that are in A after some positive iteration, and recurrence is precisely the statement that the set of points in A that are not in A after some positive iteration is a set of measure zero.

Now suppose that T is recurrent and let A be a set of positive measure. Then, $\mu(A \setminus \bigcup_{k=1}^{\infty} [A \cap T^{-k}(A)]) = \mu(A \setminus \bigcup_{k=1}^{\infty} T^{-k}(A)) = 0$. As A has positive measure, it must be the case that for some integer $n > 0$, $\mu(A \cap T^{-n}(A)) > 0$.

For the converse let A be a set of positive measure. Set

$$B = A \setminus \bigcup_{k=1}^{\infty} T^{-k}(A).$$

If $\mu(B) > 0$, then there is an integer $n > 0$ such that $\mu(B \cap T^{-n}(B)) > 0$. This would mean that there is a point $x \in B$ such that $T^n(x) \in B$, contradicting the definition of B. Therefore $\mu(B) = 0$ and T is recurrent. □

The following is probably the first theorem in ergodic theory; it was shown by Poincaré in 1899 and used in his study of celestial mechanics.

Theorem 3.5.3 (Poincaré Recurrence). *Let $(X, \mathcal{S}, \mu)$ be a finite measure space. If $T : X \to X$ is a measure-preserving transformation, then T is a recurrent transformation.*

Proof. By Lemma 3.5.2 it suffices to show that for any set of positive measure A, there is an integer $n > 0$ such that $\mu(A \cap T^{-n}(A)) > 0$. Suppose, on the contrary, that $\mu(A \cap T^{-n}(A)) = 0$ for all $n > 0$. Then for any nonnegative integers $k \ne \ell$, writing $\ell = n + k, n > 0$,

$$\mu(T^{-\ell}(A) \cap T^{-k}A) = \mu(T^{-n-k}(A) \cap T^{-k}(A))$$
$$= \mu(T^{-k}[T^{-n}(A) \cap A]) = \mu(T^{-n}(A) \cap A) = 0.$$

This means that the sets $\{T^{-n}(A)\}_{n \geq 0}$ are almost pairwise disjoint, so (see Exercise 2.4.2)

$$\mu(\bigcup_{n=0}^{\infty} T^{-n}(A)) = \sum_{n=0}^{\infty} \mu(T^{-n}(A)) = \sum_{n=0}^{\infty} \mu(A) = \infty,$$

a contradiction as $\mu(X) < \infty$. Thus it must be that $\mu(A \cap T^{-n}(A)) > 0$ for some $n > 0$. □

We present another useful characterization of recurrence. A measureable set C is said to be **compressible** for a measure-preserving transformation T if $T^{-1}(C) \subset C$ and $\mu(C \setminus T^{-1}(C)) > 0$. T is **incompressible** if it admits no compressible sets.

3.5. Recurrence

Lemma 3.5.4. *Let $(X, \mathcal{S}, \mu, T)$ be a σ-finite measure space. A measure-preserving transformation T on X is recurrent if and only if it is incompressible.*

Proof. Suppose T is recurrent and let C be a set such $T^{-1}(C) \subset C$. Put $A = C \setminus T^{-1}(C)$. We claim that

$$T^{-n}(A) \cap A = \emptyset \text{ for all } n \geq 1.$$

For suppose that $x \in A$. Then $x \in C$ and $T(x) \notin C$. So $T(x) \notin T^{-1}(C)$, or $T^2(x) \notin C$; in this way one obtains that $T^n(x) \notin C$ for all $n \geq 1$. Thus $T^n(x) \notin A$ and the claim is proved. Lemma 3.5.2 implies that $\mu(A) = 0$, so T is incompressible.

Now let T be incompressible and let A be a set of positive measure. Put $C = \bigcup_{i=0}^{\infty} T^{-i}(A)$. Then $T^{-1}(C) \subset C$. One can verify that

$$C \setminus T^{-1}(C) = A \setminus \bigcup_{i=1}^{\infty} T^{-i}(A).$$

As T is incompressible, $\mu(A \setminus \bigcup_{i=1}^{\infty} T^{-i}(A)) = 0$, which implies that T is recurrent. $\square$

Recurrence was generalized in a significant way in 1977 by Furstenberg. A measure-preserving transformation of a probability space $(X, \mathcal{S}, \mu)$ is said to be **multiply recurrent** if for all sets A of positive measure and for each integer $k > 0$ there exists an integer $n > 0$ so that

$$\mu(A \cap T^{-n}(A) \cap T^{-2n}(A) \cap \cdots \cap T^{-(k-1)n}(A)) > 0.$$

The following theorem is due to Furstenberg. Its proof, which we omit, is significantly harder than the proof of Poincaré recurrence.

Theorem 3.5.5 (Furstenberg Multiple Recurrence). *If T is a measure-preserving transformation on a finite Lebesgue measure space $(X, \mathcal{S}, \mu)$, then T is a multiply recurrent transformation.*

For a proof of this theorem, which uses techniques beyond those developed in this book, the reader is referred to [**24**]. Furstenberg used the Multiple Recurrence theorem to give a dramatically new proof of a deep result in number theory called Szemerédi's theorem. Szemerédi's theorem asserts that a set of positive density in the integers (for example, the even numbers have density 1/2) contains arithmetic progressions of arbitrary length.

Exercises

(1) Let $T : X \to X$ be a finite measure-preserving transformation. Given a measurable set A of positive measure, let n be the first integer such that $\mu(T^{-n}(A) \cap A) > 0$. Find the best upper bound (in terms of the measure of A and X) for n. Prove your claim.

(2) Let A be the set of all $x \in [0,1]$ with the following property: if $0.x_1 x_2 \ldots x_k \ldots$ is the decimal expansion of x, then for each integer $k > 0$, the string $.x_1 x_2 \ldots x_k$ appears infinitely often in the sequence $x_1 x_2 \ldots x_k \ldots$ of the decimal expansion of x. Show that $[0,1] \setminus A$ is a null set.

(3) Show that an irrational rotation is multiply recurrent without using Theorem 3.5.5.

(4) A set $P \subset \mathbb{N}$ is said to be a **Poincaré sequence** if for every finite measure-preserving system $(X, \mathcal{S}, \mu, T)$ and any set $A \in \mathcal{S}$ of positive measure there exists $n \in P, n \neq 0$, such that $\mu(T^{-n}(A) \cap A) > 0$. Show that for any infinite set P the set of differences $P - P$ (consisting of points of the form $a - b$ where $a, b \in P$) is a Poincaré sequence.

(5) A set $Q \subset \mathbb{N}$ is said to be a **thick set** if it contains intervals of integers of arbitrary length. Show that a thick set is a Poincaré sequence.

(6) A measurable set W is said to be **wandering** for a measure-preserving transformation T if for all $i, j \geq 0$ with $i \neq j$, $T^{-i}(W) \cap T^{-j}(W) = \emptyset$. Show that T is recurrent if and only if it has no wandering sets of positive measure.

3.6. Almost Everywhere and Invariant Sets

The principal notions in measure theory and ergodic theory remain invariant under a change by a set of measure zero. For example, it follows from Lemma 4.2.6 that if $T : X \to X$ is a measurable

3.6. Almost Everywhere and Invariant Sets

transformation and $S : X \to X$ is a transformation that differs from T on a null set, i.e., $\mu(\{x : T(x) \neq S(x)\}) = 0$, then S is also a measurable transformation. We shall also see in Lemma 4.6.2 that the Lebesgue integral of a function does not change if the function is changed on a set of measure zero. In addition, there are several important theorems, such as the Ergodic Theorem, that hold only after discarding a set of measure zero. The idea of carefully discarding a set of measure zero is a fundamental idea in measure theory and ergodic theory. Typically, one only needs to define objects up to a set of measure zero. We shall say that an equality holds **almost everywhere**, written as **a.e.**, if it holds outside a null set.

We start by generalizing some of the concepts we have already defined to allow changes on a null set. The first one is the notion of an invertible measurable transformation. According to the principle of ignoring things that happen on a null set, if a transformation fails to be invertible on a null set, it should still be considered invertible. So one is led to define the notion of an invertible transformation mod μ as a transformation that is invertible after discarding a μ-measure zero set from its domain. However, the remaining map should be a transformation, i.e., points that are not in the null set should also miss the null set under iteration by the transformation. More precisely, if $T : X \to X$ is a transformation and $N \subset X$ is the null set that is discarded, it is necessary that if $x \in X \setminus N$, then $T(x) \in X \setminus N$. To express this idea in a clear way we are led to the notion of an *invariant set*, which plays a crucial role in dynamics.

Let $T : X \to X$ be a transformation. A subset $A \subset X$ is said to be **positively invariant** or **positively T-invariant** when

$$x \in A \text{ implies } T(x) \in A.$$

Equivalently, $A \subset T^{-1}(A)$. Then, the transformation T restricted to A defines a transformation $T : A \to A$. It is also clear that if T restricted to A defines a transformation, then A is a positively invariant set. In some cases we need an additional property: A set A is **strictly T-invariant** when $T^{-1}(A) = A$, i.e., when $x \in A$ if and only if $T(x) \in A$. In ergodic theory we often call a strictly invariant set simply **invariant** or **T-invariant**.

Example. Let $X = [0, \infty)$ and define $T : X \to X$ by $T(x) = x + 1$, an invertible measure-preserving transformation. Then $A = [1, \infty)$ is positively invariant but not strictly invariant.

Question. Let $T : X \to X$ be a transformation. Show that the empty set and the whole space X are positively invariant sets. Show that X is strictly invariant if and only if T is onto.

Example. Let $X = \{0, 1, 2, \ldots, 5\}$. Define $T : X \to X$ by $T(0) = 1, T(1) = 2, T(2) = 0, T(3) = 4, T(4) = 5, T(5) = 3$, and $S(i) = i + 1$ for $i = 0, \ldots, 4$, $S(5) = 0$. Let $A = \{0, 1, 2\}$. Then A is a strictly T-invariant set but is not a positively S-invariant set. T restricted to A is a rotation on three points. One can also verify that S does not have any positively invariant sets other than $\emptyset$ and X.

Given two sets $A, B \subset X$, we say that

$$A = B \bmod \mu \text{ if } \mu(A \triangle B) = 0.$$

(Recall that $A \triangle B = \emptyset$ if and only if $A = B$.) A measurable set A is said to be **strictly invariant mod** μ (or **T-invariant mod** μ) if $A = T^{-1}(A) \bmod \mu$.

Question. Let $(X, \mathcal{S}, \mu)$ be a measure space and $T : X \to X$ a transformation. Show that if T is measure-preserving, then X is strictly invariant mod μ.

The following lemma is an immediate consequence of Lemma 3.5.4 but it is important to keep in mind.

Lemma 3.6.1. *Let T be a recurrent measure-preserving transformation on a measure space $(X, \mathcal{S}, \mu)$. If A is positively invariant, then it is strictly invariant mod μ.*

Furthermore, Lemma 3.6.2 shows that a strictly invariant mod μ set differs from a strictly invariant set in a null set. Therefore, in the case of finite measure-preserving transformations, as we identify sets that differ in a null set, a positively invariant set may be replaced by a strictly invariant set; this means that there is no essential distinction between positively invariant sets and strictly invariant sets. We shall

3.6. Almost Everywhere and Invariant Sets

see some examples from topological dynamics and infinite measure-preserving transformations where the distinction is important. For example, it can be shown that the transformation $T : \mathbb{Z} \to \mathbb{Z}$ defined on the integers with counting measure by $T(n) = n + 1$ is an infinite measure-preserving transformation that has no strictly invariant sets of positive measure other than $\mathbb{Z}$, but the set $A = \{n \in \mathbb{Z} : n > 0\}$ is an invariant set for T. There are also examples on nonatomic spaces when T is not invertible.

Lemma 3.6.2. Let $(X, \mathcal{S}, \mu)$ be a σ-finite measure space and let $T : X \to X$ be a measure-preserving transformation. If $A \in \mathcal{S}$ is strictly T-invariant mod μ, then there exists a set $\widehat{A}$ that differs from A by a null set (i.e., $\mu(A \triangle \widehat{A}) = 0$) that is strictly T-invariant.

Proof. We assume that μ is finite and leave the σ-finite case to the reader. As A is strictly T-invariant mod μ, $\mu(T^{-1}(A) \triangle A) = 0$. By the triangle inequality (see Exercise 2.5.6) and the fact that T is measure-preserving,

$$\mu(T^{-2}(A) \triangle A) \leq \mu(T^{-2}(A) \triangle T^{-1}(A)) + \mu(T^{-1}(A) \triangle A) = 0.$$

So by induction,

$$\mu(T^{-n}(A) \triangle A) = 0 \text{ for all } n \geq 0.$$

By Exercise 4, the set $\widehat{A} = \bigcap_{k=1}^{\infty} T^{-k}(A^+)$, where $A^+ = \bigcup_{n=1}^{\infty} T^{-n}(A)$, is strictly T-invariant. Now

$$\mu(A \triangle T^{-k}(A^+)) = \mu(A \triangle \bigcup_{n=k+1}^{\infty} T^{-n}(A))$$

$$\leq \sum_{n=k+1}^{\infty} \mu(A \triangle T^{-n}(A)) = 0.$$

Then by Theorem 2.5.2b) (here is where we use that μ is finite),

$$\mu(A \triangle \bigcap_{k=1}^{\infty} T^{-k}(A^+)) = 0,$$

so $\mu(A \triangle \widehat{A}) = 0$. □

We conclude by applying the principle of neglecting sets of measure zero to extend the notion of an invertible transformation.

A measurable transformation $T : X \to X$ is said to be an **invertible measurable transformation** mod μ (or mod 0 when the measure is clear from the context) if there exists a measurable set $X_0 \subset X$, with $\mu(X \setminus X_0) = 0$ and such that T is one-to-one on X_0 (i.e., for any x_1, x_2 in X_0, if $T(x_1) = T(x_2)$, then $x_1 = x_2$), $T(X_0) = X_0$ and $T^{-1} : X_0 \to X_0$ is a measurable transformation (it follows that T^{-1} is measure-preserving when T is measure-preserving). In the literature one may see authors refer to "invertible transformations" when what is really meant is "invertible transformations mod μ." The idea here is that after discarding a set of measure zero one obtains an invertible transformation.

Exercises

(1) Let T be an invertible mod μ measurable transformation. Show that T is measure-preserving if and only if for all measurable sets A, $A \subset X_0$, $T(A)$ is measurable and $\mu(T(A)) = \mu(A)$.

(2) Complete the proof of Lemma 3.6.2 in the σ-finite case.

We define some sets that are useful in discussing invariance properties and are used in the following exercises. Given a transformation $T : X \to X$ and a set $A \subset X$, define the sets

$$A^+ = \bigcup_{n=1}^{\infty} T^{-n}(A); \quad A^{\oplus} = \bigcup_{n=0}^{\infty} T^{-n}(A).$$

A^+ represents the set of points in X that at some time $n > 0$ enter A. For example, $\mu(A^+ \cap A) > 0$ means that there is a set of positive measure of points of A that "come back" to A at some positive time. The set $A^{\oplus}$ is precisely A^+ with the addition of A. We note, however, that A^+ is not positively invariant, but as $T^{-1}(A^+) \subset A^+$, it can be shown that the set

$$\widehat{A} = \bigcap_{n=1}^{\infty} T^{-n}(A^+)$$

3.7. Ergodic Transformations

is strictly invariant. When T is invertible it is often more convenient to use the set

$$A^\star = \bigcup_{n=-\infty}^{\infty} T^{-n}(A).$$

(3) Let $T : X \to X$ be a nonsingular (see p. 152) transformation and let $N \subset X$. Show that if N is a null set, then $N^\oplus$ is a null set and $X \setminus N^\oplus$ is a positively T-invariant set, of the same measure as X, which does not contain N. Furthermore, if $T : X \to X$ is measure-preserving, then $T : X \setminus N^\oplus \to X \setminus N^\oplus$ is measure-preserving.

(4) Let T be a nonsingular transformation (see p. 152). Show that the set $\widehat{A}$ is strictly T-invariant and satisfies

$$\widehat{A} = \bigcap_{n=k}^{\infty} T^{-n}(A^+) = \bigcap_{n=k}^{\infty} T^{-n}(A^\oplus)$$

for all $k > 0$. If A has measure zero, then $\widehat{A}$ has measure zero. Furthermore, if T is measure-preserving and A has finite measure, then $\mu(A) = \mu(\widehat{A})$.

(5) Show that if T is invertible, then $A^\star$ is a strictly T-invariant set containing A. If furthermore N is a null set, then $N^\star \supset N$ is a null set and $T : X \setminus N^\star \to X \setminus N^\star$ is an invertible transformation, and if $T : X \to X$ is measure-preserving, then $T : X \setminus N^\star \to X \setminus N^\star$ is measure-preserving.

3.7. Ergodic Transformations

Ergodicity is one of our most important concepts. Originally called *metric transitivity*, it was introduced by Birkhoff and Smith in 1928. We will see that there are several equivalent formulations of ergodicity, each having its own interesting interpretation. One of the equivalences is surprising and the result of a deep theorem, namely the ergodic theorem (Theorem 5.1.1). We will discuss that interpretation later. In this section we study the basic definition of ergodicity and show that irrational rotations and the doubling map are ergodic transformations.

We are interested in studying the measurable dynamics of a transformation T defined on a space X. If there existed a set $A \subset X$ such that $x \in A$ if and only if $T(x) \in A$, then T restricted to A (i.e., as a map $T : A \to A$) and T restricted to A^c would both be dynamical systems in their own right, and it is reasonable to think that the study of the dynamics of T could be reduced to the study of the dynamics of T restricted to A and T restricted to A^c. Thus one can think of systems not having such strictly invariant sets A (other than $\emptyset$ and X) as "indecomposable" or basic systems.

A measure-preserving transformation T is said to be **ergodic** if whenever A is a strictly T-invariant measurable set, then either $\mu(A) = 0$ or $\mu(A^c) = 0$. The following is the simplest example of an ergodic transformation.

Example. Let $(X, \mathcal{S}, \mu)$ be a one-point canonical Lebesgue space (i.e., $X = \{p\}$, $\mathcal{S} = \{\emptyset, X\}$ and μ is the counting measure on X defined by $\mu(\{p\}) = 1$). Let T be the identity transformation on X (i.e., $T(p) = p$). Then T is a measure-preserving ergodic transformation. To see that it is ergodic note that the strictly invariant sets are $\emptyset$ and $\{p\}$, and they have measure 0 or their complement has measure 0.

Question. Let T be the identity transformation on a canonical Lebesgue measure space X. Show that T is ergodic if and only if X is a one-point space.

Question. Show that the transformation $R(x) = x+3/4$ (mod 1) is a finite measure-preserving transformation on $[0, 1)$ that is not ergodic.

The following lemma shows that in the definition of ergodicity one may consider strictly invariant sets up to measure and not just strictly invariant.

Lemma 3.7.1. *Let $(X, \mathcal{S}, \mu)$ be a σ-finite measure space and let $T : X \to X$ be a measure-preserving transformation. Then T is ergodic if and only if when A is strictly invariant $\operatorname{mod} \mu$, then $\mu(A) = 0$ or $\mu(A^c) = 0$.*

Proof. This is a direct consequence of Lemma 3.6.2. □

3.7. Ergodic Transformations

The transformation $T : \mathbb{Z} \to \mathbb{Z}$ defined by $T(n) = n + 1$, with counting measure on $\mathbb{Z}$, is ergodic but not recurrent. As we have seen, recurrence always holds in the finite measure-preserving case, and most interesting characterizations of ergodicity are in the recurrent case.

Lemma 3.7.2. *Let $(X, \mathcal{S}, \mu)$ be a σ-finite measure space and let $T : X \to X$ be a measure-preserving transformation. Then the following are equivalent.*

(1) T *is recurrent and ergodic.*

(2) *For every set A of positive measure, $\mu(X \setminus \bigcup_{n=1}^{\infty} T^{-n}(A)) = 0$. (In this case we will say A **sweeps out** X.)*

(3) *For every measurable set A of positive measure and for a.e. $x \in X$ there exists an integer $n > 0$ such that*
$$T^n(x) \in A.$$
(This is illustrated in Figure 3.8.)

(4) *If A and B are sets of positive measure, then there exists an integer $n > 0$ such that*
$$T^{-n}(A) \cap B \neq \emptyset.$$

(5) *If A and B are sets of positive measure, then there exists an integer $n > 0$ such that*
$$\mu(T^{-n}(A) \cap B) > 0.$$

Proof. First we show that (5) implies (1). Let A be a strictly invariant set of positive measure. Then $T^{-n}(A) = A$ for all $n > 0$. Let $B = A^c$. If B had positive measure, then there would exist an integer $n > 0$ so that $\mu(T^{-n}(A) \cap B) > 0$, so $\mu(A \cap A^c) > 0$, a contradiction. Therefore, $\mu(A^c) = 0$, which means that T is ergodic.

Suppose (1) holds. Let A be a set of positive measure and set $A^+ = \bigcup_{n=1}^{\infty} T^{-n}(A)$. Then $T^{-1}(A^+) \subset A^+$; since T is recurrent $\mu(A^+ \setminus T^{-1}(A^+)) = 0$. Therefore A^+ is strictly invariant mod μ and since it has positive measure the ergodicity of T implies that $A^+ = X$ mod μ.

It is clear that (2) is equivalent to (3). Now assume (2) and let A and B be sets of positive measure. Then $A^+ = X \mod \mu$ and so there is $n > 0$ with $\mu(T^{-n}(A) \cap B) > 0$. Therefore (4) holds.

Assume (4). Let A, B be sets of positive measure and suppose $\mu(T^{-n}(A) \cap B) = 0$ for all $n > 0$. Then put $A_0 = A \setminus \bigcup_{n=1}^{\infty} T^{-n}(A) \cap B$. Then $\mu(A_0) > 0$ but $T^{-n}(A_0) \cap B = \emptyset$, contradicting (4).

Now if (5) is true and A is strictly T-invariant (then $T^{-n}(A) = A$ for all $n > 0$), using $B = A^c$ in (5) we see that A and A^c cannot both have positive measure. Therefore, T is ergodic. Recurrence follows by letting $B = A$. $\square$

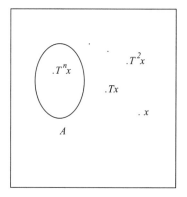

Figure 3.8. T is recurrent and ergodic

We now introduce some approximation techniques that will prove useful in showing ergodicity, and apply them to the case of irrational rotations and the doubling map.

Given measurable sets A and I (I will typically be in some sufficient semi-ring, i.e., an interval or a dyadic interval) and given $1 > \delta > 0$ (usually chosen small), we say that I is $(1-\delta)$-**full** of A if
$$\lambda(A \cap I) > (1-\delta)\lambda(I).$$
While we say "$(1-\delta)$-full," strictly speaking we should say "more than $(1-\delta)$-full." If $\delta = 1/4$, say, then "I is $3/4$-full of A" can be interpreted as saying that "in measure, more than $3/4$ of I is in A."

3.7. Ergodic Transformations

The following lemma shows that for any set of positive finite measure and any $\delta > 0$ there is always an element of a sufficient semi-ring that is $(1-\delta)$-full of the set.

Lemma 3.7.3. *Let $(X, \mathfrak{L}, \lambda)$ be a measure space with a sufficient semi-ring $\mathcal{C}$. If $A \in \mathfrak{L}$ is of finite positive measure, then for any $1 > \delta > 0$ there exists $I \in \mathcal{C}$ such that I is $(1-\delta)$-full of A.*

Proof. Let $\varepsilon > 0$ be such that $\varepsilon < \frac{\delta}{2-\delta}$ (this choice of ε will be apparent later in the proof). By Lemma 2.7.3 there exists $H^* = \bigsqcup_{j=1}^{N} I_j$, with the sets I_j disjoint and in $\mathcal{C}$, such that $\lambda(A \triangle H^*) < \varepsilon \lambda(A)$. By Exercise 2.5.4,

$$\lambda(A \cap H^*) > (1-\varepsilon)\lambda(A).$$

Also, by Exercise 2.5.5,

$$\lambda(H^*) < \lambda(A) + \varepsilon\lambda(A) = (1+\varepsilon)\lambda(A).$$

Now assume that for all $j \in \{1, \ldots, N\}$,

$$\lambda(A \cap I_j) \leq (1-\delta)\lambda(I_j).$$

Then $\lambda(A \cap H^*) \leq (1-\delta)\lambda(H^*)$. Therefore,

$$\lambda(A \cap H^*) < (1-\delta)(1+\varepsilon)\lambda(A).$$

Now ε was chosen so that $(1-\delta)(1+\varepsilon) < 1-\varepsilon$, but this means that $\lambda(A \cap H^*) < \lambda(A \cap H^*)$, a contradiction. Therefore there must exist $j \in \{1, \ldots, N\}$ such that $\lambda(A \cap I_j) > (1-\delta)\lambda(I_j)$. □

The following lemma contains a basic inequality that will be used in several proofs of ergodicity. It follows directly from set theory.

Lemma 3.7.4. *Let T be a measure-preserving transformation and let A, B, I, J be any measurable sets such that $A \subset I$ and $B \subset J$. Then for all integers $n \geq 0$,*

(3.6) $\quad \lambda(T^{-n}(A) \cap B) \geq \lambda(T^{-n}(I) \cap J) - \lambda(I \setminus A) - \lambda(J \setminus B).$

Proof. Note that

$$T^{-n}(I) \cap J \subset (T^{-n}(A) \cap B) \cup (T^{-n}(I) \setminus T^{-n}(A)) \cup (J \setminus B),$$

and $\lambda(T^{-n}(I) \setminus T^{-n}(A)) = \lambda(I \setminus A)$. (See Figure 3.9.) □

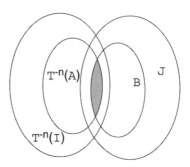

Figure 3.9. $T^{-n}(A) \cap B$

We are now ready for our main result.

Theorem 3.7.5. *Irrational rotations are ergodic.*

Proof. Let R be the rotation by the irrational number α on $X = [0,1)$. Let A_1 and B_1 be any sets of positive measure. Then, there exist dyadic intervals I and J such that

$$\lambda(A_1 \cap I) > \frac{3}{4}\lambda(I) \quad \text{and} \quad \lambda(B_1 \cap J) > \frac{3}{4}\lambda(J).$$

Furthermore, we may assume that I and J are of the same measure (if J, say, is bigger than I, then at least one of the two halves of J must be $\frac{3}{4}$-full of B_1; continue in this way until obtaining a subinterval of J of the same measure as I that is $\frac{3}{4}$-full of B_1, finally rename it J). Write

$$A = A_1 \cap I \text{ and } B = B_1 \cap J.$$

Suppose $I = [a,b)$ is to the left of $J = [c,d)$ in $[0,1)$, i.e., $a \leq c$. As the orbit of b under R is dense, there is an integer $n > 0$ such that

$$d - \frac{d-c}{4} < R^n(b) < d.$$

Therefore $\lambda(R^n(I) \cap J) > \frac{3}{4}\lambda(J)$. Thus by Lemma 3.7.4,

$$\lambda(R^n(A) \cap B) \geq \lambda(R^n(I) \cap J) - \lambda(I \setminus A) - \lambda(J \setminus B)$$
$$> \frac{3}{4}\lambda(J) - \frac{1}{4}\lambda(I) - \frac{1}{4}\lambda(J) > 0.$$

Therefore R is ergodic. □

3.7. Ergodic Transformations

A measure-preserving transformation T is said to be **totally ergodic** if for all integers $n > 0$, T^n is ergodic. Irrational rotations are our first example of totally ergodic transformations as $R_\alpha^n = R_{n\alpha}$ and so R_α^n is ergodic for all $n \neq 0$ if α is irrational.

We end with an example of a noninvertible ergodic transformation.

Theorem 3.7.6. *Let $T(x) = 2x \,(\mathrm{mod}\, 1)$ be defined on $[0,1)$. Then T is an ergodic finite measure-preserving transformation on $[0,1)$.*

Proof. We have already seen that T is finite measure-preserving.

To show ergodicity, first let $D_{n,k} = [\frac{k}{2^n}, \frac{k+1}{2^n})$ ($n > 0$, $k = 0, \ldots, 2^n - 1$) be a dyadic interval in $[0,1)$. We note that $T^n(D_{n,k}) = [0,1)$. Using this, it follows by induction that $T^{-n}(D_{n,k})$ consists of 2^n disjoint dyadic intervals each of length 2^{-2n}. In Exercise 4, the reader is asked to show by induction on n that for any measurable set A,

$$\lambda(T^{-n}(A) \cap D_{n,k}) = \frac{1}{2^n}\lambda(A) = \lambda(A)\lambda(D_{n,k})$$

for $k = 0, \ldots, 2^n - 1$. Suppose now that A is a strictly T-invariant set. So $T^{-n}(A) = A$ for all $n > 0$, and

$$\lambda(A \cap D_{n,k}) = \lambda(A)\lambda(D_{n,k}).$$

If A has positive measure, as the dyadic intervals form a sufficient semi-ring, for any $\delta > 0$ there exists a dyadic interval $D_{n,k}$ (for some n, k) so that $\lambda(A \cap D_{n,k}) > (1-\delta)\lambda(D_{n,k})$. As δ is arbitrary, this implies that $\lambda(A) = 1$ and therefore T is ergodic. $\square$

Exercises

(1) Show that if R is an irrational rotation, then R^n is ergodic for all $n \neq 0$.

(2) Let T be a rotation on n points, i.e., $X = \{a_0, a_1, \ldots, a_{n-1}\}$, μ a measure on subsets of X defined by $\mu(\{a_i\}) = 1/n$, and T a transformation on X defined by $T(a_i) = a_{i+1} \,(\mathrm{mod}\, n)$. Show that T is ergodic.

(3) Let T be a totally ergodic measure-preserving transformation. Show that if T is invertible, then T^n is ergodic for all $n < 0$.

(4) Complete the details in the proof of Theorem 3.7.6.

(5) Show that for each integer $k > 1$ the transformation $T(x) = kx \pmod 1$ is ergodic on $[0, 1)$.

(6) Show that the 2-dimensional baker's transformation is an invertible (mod λ^2) measure-preserving ergodic transformation.

(7) Let T be the doubling map of Theorem 3.7.6. Is T totally ergodic?

3.8. The Dyadic Odometer

The transformation studied in this section was defined by Kakutani and von Neumann in the 1940's and is also called the Kakutani–von Neumann odometer or the dyadic adding machine. We first present its definition as a piecewise-linear map on infinitely many pieces in the unit interval. While this defines the transformation on the unit interval and helps understand why it is measure-preserving, there is another definition that is better for discussing its dynamical properties, such as ergodicity. This other definition also introduces a new method for constructing transformations, called *cutting and stacking*, which will provide us with a wealth of examples and counterexamples.

We start by defining $T : [0, 1) \to [0, 1)$ by

$$T(x) = \begin{cases} x + \frac{1}{2}, & \text{if } 0 \leq x < 1/2; \\ x - \frac{5}{8}, & \text{if } 1/2 \leq x < 3/4; \\ x - \frac{1}{4}, & \text{if } 3/4 \leq x < 7/8; \\ \vdots \end{cases}$$

The inductive process should be clear from the definition of T and its partial graph in Figure 3.10: the unit interval is subdivided into a countable number of abutting intervals, starting with $[0, 1/2)$, so that the ith interval has length $1/2^i$. Then T sends $[0, 1/2)$ to $[1/2, 1)$ by the translation map $x \to x + 1/2$, and the remaining intervals

3.8. The Dyadic Odometer

are sent by the corresponding translation to the interval of the same length preceding the previous interval. So, for example, the interval $[1/2, 3/4)$ is sent to the interval $[1/4, 1/2)$ (the interval of length $1/4$ preceding $[1/2, 1)$).

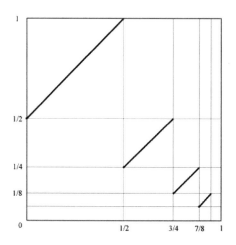

Figure 3.10. Graph of the odometer map

Now we introduce another description of this map. First we need some notation. Given two intervals $I = [a, b)$ and $J = [c, d)$ of the same length, there is a unique orientation-preserving translation from I to J, namely the map $T_{I,J}$ defined by $T_{I,J}(x) = x + c - a$, sending the left point of I to the left point of J. The main properties that we use of $T_{I,J}$ are:

(1) $T_{I,J}$ is determined by I and J and is a one-to-one map from I onto J;

(2) for any measurable set $A \subset I$, $T_{I,J}(A) \subset J$ is measurable and $\lambda(T_{I,J}(A)) = \lambda(A)$ (as Lebesgue measure is translation invariant);

(3) if I' and J' are dyadic subintervals of I and J, respectively, of the same order, then $T_{I,J}$ agrees with $T_{I',J'}$ on I'.

In the diagrams, the action of $T_{I,J}$ on I will be denoted by an arrow as in Figure 3.11.

Figure 3.11. Map $T_{I,J}$

With this notation we can describe the value of T on $[0, 1/2)$ by giving two intervals, the intervals $[0, 1/2)$ and $[1/2, 1)$, with the interval $[1/2, 1)$ placed above the interval $[0, 1/2)$ as in Figure 3.12.

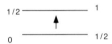

Figure 3.12. The odometer on $[0, 1/2)$

To completely define the transformation T we specify a process that generates a sequence of *columns*. A **column**, sometimes also called a **tower**, consists of a finite sequence of disjoint intervals of the same length. The intervals in a column serve to specify the value of the transformation on each interval except the top. Each interval in a column is called a **level**. The number of levels in a column is called the **height** of the column.

The first column will consist of the unit interval, so

$$C_0 = \{[0, 1)\}.$$

(All our intervals will be left-closed and right-open.) We describe how to obtain the next column. It is useful to think of this in terms of cutting and stacking the levels of the previous column. To obtain column C_1, cut, i.e., divide, the single level in C_0 into the two disjoint subintervals $[0, 1/2)$ and $[1/2, 1)$ and stack them so that the right subinterval is placed above the left subinterval to form column $C_1 = \{[0, 1/2), [1/2, 1)\}$ of height $h_1 = 2$. Denote the levels of C_1, from bottom to top, by $I_{1,0}$ and $I_{1,1}$. Column C_1 defines a partial map T_{C_1} by the translation that sends $[0, 1/2)$ to $[1/2, 1)$. Note that T_{C_1} is not

3.8. The Dyadic Odometer

defined on $[1/2, 1)$. Figure 3.13 illustrates the process of "cutting and stacking" C_0 to obtain C_1 and also shows the resulting column C_1.

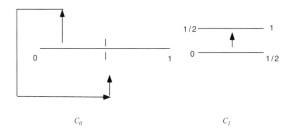

Figure 3.13. Generating column C_1

Before proceeding to the inductive step, to better clarify the construction, we show how to obtain C_2 from C_1. First, cut each level of C_1 in half, and then stack the right subcolumn above the left subcolumn to obtain the new levels $I_{2,0} = [0, 1/4)$, $I_{2,1} = [1/2, 3/4)$, $I_{2,2} = [1/4, 1/2)$ and $I_{2,3} = [3/4, 1)$. This gives the $h_2 = 4$ levels of C_2. Observe that T_{C_2} agrees with T_{C_1} wherever T_{C_1} is defined, but T_{C_2} is now defined on the left half of the top level of C_1; of course, T_{C_2} remains undefined on the top level of C_2. Figure 3.14 shows the process of obtaining C_2 from C_1, and also shows the resulting column C_2. As one can read column C_2 from the left part of Figure 3.14, this is the figure we find more useful.

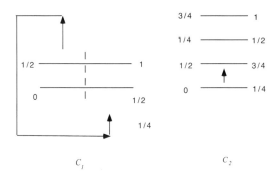

Figure 3.14. Generating column C_2

Finally, to obtain column C_{n+1} from $C_n = \{I_{n,0}, \ldots, I_{n,h_n-1}\}$, cut each level of C_n in half and stack the right subintervals above the left subintervals to obtain a column of height $h_{n+1} = 2h_n$. This is shown in Figure 3.15. One can verify that $h_n = 2^n$.

We have now defined a sequence of partial maps $\{T_{C_n}\}_{n\geq 0}$ so that $T_{C_{n+1}}$ agrees with T_{C_n} wherever T_{C_n} is defined and T_{C_n} is undefined on a subinterval of $[0,1)$ of measure $\frac{1}{2^n}$ (namely, the top level of C_n). Define the dyadic odometer T by $T(x) = \lim_{n\to\infty} T_{C_n}(x)$. For each $x \in [0,1)$ there is some integer $n > 0$ so that x belongs to some level of C_n that is not the top level. This means that $T(x)$ is well defined. A similar argument shows that T is a one-to-one transformation of $X = [0,1)$. Furthermore, $T^{-1}(x)$ is defined for all $x \in (0,1)$. Note that $T^{-1}(0)$ is not defined, but it is not hard to see that T is invertible mod λ. In fact, define X_0 by $X_0 = X \setminus \bigcup_{n=0}^{\infty}\{T^n(0)\}$; in other words, delete from X the endpoints of all the dyadic intervals. Then, $\lambda(X_0) = 1$ and now $T : X_0 \to X_0$ is one-to-one and onto.

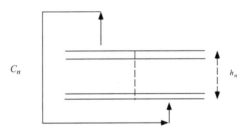

Figure 3.15. Column C_n for the dyadic odometer

Let $n > 0$ and let I be a level in C_n. We introduce some notation to label subintervals in I. When constructing C_{n+1} from C_n, each level in C_n is subdivided into two left-closed right-open, disjoint subintervals that are levels in C_{n+1}; call the first (leftmost) one $I[0]$ and the second one $I[1]$. So, $I_{n,i}[0]$ is $I_{n+1,i}$ and $I_{n,i}[1]$ is $I_{n+1,i+h_n}$. This notation is extended inductively in the following way. If $I[a_0a_1\cdots a_{k-1}]$ (a level in C_{n+k}) has been defined, let $I[a_0a_1\cdots a_{k-1}0]$ be the left sublevel of $I[a_0a_1\cdots a_{k-1}]$ in C_{n+k+1}, and similarly let $I[a_0a_1\ldots a_{k-1}1]$ be the right sublevel of $I[a_0a_1\cdots a_{k-1}]$ in C_{n+k+1}.

3.8. The Dyadic Odometer

Lemma 3.8.1. *Let T be the dyadic odometer. Then,*

(1) *for all $n > 0$, $T(I_{n,h_n-1}) = I_{n,0} \setminus \{0\}$,*

(2) *for all $n > 0$, $i = 0, \ldots, h_n - 1$,*

$$T^{h_k}(I_{n,i}) = I_{n,i} \setminus \{0\}, \text{ for all } k \geq n.$$

Proof. Let $I = I_{n,h_n-1}$ and $J = I_{n,0}$ be the top and bottom levels of C_n, respectively. From the definition of T_{C_n}, $T(I[0]) = J[1]$. Since this is true for all $n > 0$, using that $I[1], J[0]$ are the top and bottom, respectively, levels of C_{n+1} one obtains that $T(I[1,0]) = T((I[1])[0]) = J[0][1] = J[01]$. Similarly, $T(I[110]) = J[001]$. If we let 1^k denote k consecutive 1's, and similarly for 0^k, by induction we see that $T(I[1^k 0]) = J[0^k 1]$. As $I[1] = \bigsqcup_{k>0} I[1^k 0]$, this completes the proof of part (1). The second part is proved similarly. □

A consequence of Lemma 3.8.1 is that for each $k > 0$ the transformation T^{h_k} is not ergodic.

The proof of the following lemma is a direct application of the definition and left to the reader. A more general version will be proved later (Lemma 6.5.4).

Lemma 3.8.2. *Let A be a set of positive measure and let I be a dyadic interval that is $3/4$-full of A. Let I_0 and I_1 be the first and second half subintervals of I, respectively. Then one of I_0 or I_1 is $3/4$-full of A and both I_0 and I_1 are $1/2$-full of A.*

Theorem 3.8.3. *If T is the dyadic odometer, then T is a measure-preserving and ergodic invertible mod λ transformation of $[0,1)$.*

Proof. Since the dyadic intervals form a sufficient semi-ring and T is measurable and measure-preserving on the dyadic intervals, Theorem 3.4.1 implies that T is measurable and measure-preserving. That T is invertible mod λ follows from the existence of the set X_0 mentioned above.

To show that T is ergodic let A_1 and B_1 be two sets of positive measure in $[0,1)$. There exist dyadic intervals I and J such that I and J are $\frac{3}{4}$-full of A_1 and B_1, respectively. As before, we may assume that I and J are of the same measure. Thus they are both levels in some column C_{m-1}, say, for some $m > 0$. As each half of J in column C_m is at least $\frac{1}{2}$-full of B_1, by considering the appropriate subintervals in column C_m we may finally assume that we have intervals I and J in C_m, each $\frac{1}{2}$-full of A_1 and B_1, respectively, and with J above I. Let $A = A_1 \cap I$ and $B = B_1 \cap J$. There is an integer $n > 0$ such that

$$T^n(I) = J \mod \lambda.$$

Therefore,

$$\begin{aligned}\lambda(T^n(A_1) \cap B_1) &\geq \lambda(T^n(A) \cap B) \\ &\geq \lambda(T^n(I) \cap J) - \lambda(I \setminus A) - \lambda(J \setminus B) \\ &> \lambda(J) - \frac{1}{2}\lambda(I) - \frac{1}{2}\lambda(J) = 0.\end{aligned}$$

Thus the transformation is ergodic. □

The construction of the dyadic odometer can be generalized in a natural way. Let $\{r_n\}_{n \geq 0}$ be a sequence of integers ≥ 2. We let $\{r_n\}_{n \geq 0}$ determine a sequence of columns $\{C_n\}_{n \geq 0}$ in the following way. Let $C_0 = \{[0,1)\}$. Assuming that C_n has been defined, let C_{n+1} be the column obtained from C_n by cutting each level of C_n into r_n equal-length subintervals and stacking from left to right. This defines a new transformation T called the r_n-**odometer** with $h_{n+1} = r_n h_n$. For example, the dyadic odometer is obtained when $r_n = 2$ for all $n \geq 0$. The exercises explore properties of these transformations.

Exercises

(1) Give another proof that the dyadic odometer is ergodic by showing directly that if A is a T-invariant set of positive measure, then $\lambda(A) = 1$.

(2) Let T be the dyadic odometer. Show that for every $n > 0$ and every level I in column C_n, $T^{h_n}(I) = I \mod \lambda$. Use this to deduce that for every set of positive measure A and every integer $k > 0$ there is an integer $n > 0$ such that

$$\lambda(A \cap T^n(A) \cap T^{2n}(A) \cap \ldots \cap T^{kn}(A)) > 0.$$

(This is Furstenberg's Multiple Recurrence property for the dyadic odometer.)

(3) Let T be the dyadic odometer. Show that for all n, the transformation T^{2n} is not ergodic. What about ergodicity of T^k for k odd?

(4) Construct a finite measure-preserving transformation T so that T and T^2 are ergodic but T^3 is not ergodic.

(5) Let T be the r_n-adic odometer with $r_n = n$ for $n \in \mathbb{N}$ (i.e., for each $n > 2$ column C_n is cut into $r_n = n$ subcolumns). Show that the transformation T^k is ergodic if and only if $k = 1$.

(6) Let T be the shift on $\mathbb{Z}$ with counting measure. Is T totally ergodic?

(7) Let T be the dyadic odometer. For each $n > 0$, extend the column map T_{C_n} so that it is defined on the top level of C_n by the translation that takes the interval I_{n,h_n-1} to the interval $I_{n,0}$. (Note that while T also takes I_{n,h_n-1} to $I_{n,0}$, it is not a translation on I_{n,h_n-1} and differs from T_{C_n} on points of this interval.) a) Show that the extended map T_{C_n} is a nonergodic finite measure-preserving map of $[0,1)$. b) Show that the sequence of maps T_{C_n} converges to T in the sense that for every measurable set A,

$$\lim_{n \to \infty} \lambda(T_{C_n}(A) \triangle T(A)) = 0.$$

3.9. Infinite Measure-Preserving Transformations

We have seen that an infinite measure-preserving transformation does not have to be recurrent. For example, $T(x) = x + 1$ on $\mathbb{R}$ with Lebesgue measure is measure-preserving but not recurrent. Also, $T(n) = n + 1$ on $\mathbb{Z}$ with counting measure is measure-preserving and ergodic but not recurrent. We start with a lemma showing that for invertible transformations on nonatomic spaces, ergodicity implies recurrence. We then present an example of an invertible measure-preserving transformation on the positive reals that is ergodic.

Lemma 3.9.1. *Let T be an invertible measure-preserving transformation on a nonatomic σ-finite measure space $(X, \mathcal{S}, \mu)$. If T is ergodic, then it is recurrent.*

Proof. Suppose that T is not recurrent. Then there exists a set A of positive measure such that $\mu(T^{-n}(A) \cap A) = 0$ for all $n > 0$. It follows that $\mu(\bigcup_{n \neq 0}(T^n(A) \cap A)) = 0$ (here the notation means that the union is taken over all positive and negative n that are nonzero). Then the set
$$W = A \setminus (\bigcup_{n \neq 0}(T^n(A) \cap A))$$
satisfies
$$\mu(W) = \mu(A) > 0 \text{ and}$$
$$T^n(W) \cap T^m(W) = \emptyset \text{ for all } m \neq n.$$
Since $(X, \mathcal{S}, \mu)$ is nonatomic, there exists $B \subset W$ such that $0 < \mu(B) < \mu(W)$. The set $B^\star = \bigcup_{n=-\infty}^{\infty} T^n(B)$ is T-invariant and one can verify that $\mu(B^\star) > 0$ and $\mu((B^\star)^c) > 0$. This contradicts the fact that T is ergodic. Therefore T is recurrent. □

We construct now an ergodic infinite measure-preserving invertible transformation T that was introduced by Hajian and Kakutani in 1970. The construction of this transformation uses the method of cutting and stacking; we will need to inductively define a sequence of columns $\{C_n\}$. While in the case of the dyadic odometer the union of the levels of any column is always the interval $[0, 1)$, here column C_{n+1} will contain new levels that are not contained in the union of the levels of C_n; these new levels are called **spacers**. The union of the levels over all the columns will be $[0, \infty)$. Thus the transformation will be defined on an infinite measure space.

We start by letting $C_0 = \{[0, 1)\}$, $h_0 = 1$ and $X_0 = [0, 1)$. Before giving the inductive step, we explain how column C_1 is obtained from C_0. As in the case of the dyadic odometer, subdivide $[0, 1)$ into the equal length subintervals $[0, 1/2), [1/2, 1)$, but in this case we add two new subintervals of length $1/2$. We choose the subintervals abutting $[0, 1)$ so that they are $[1, 3/2)$ and $[3/2, 2)$. Then $C_1 = \{[0, 1/2), [1/2, 1), [1, 3/2), [3/2, 2)\}$; the height is $h_1 = 4h_0 = 4$.

3.9. Infinite Measure-Preserving Transformations

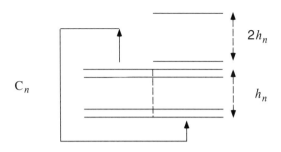

Figure 3.16. Generating C_{n+1} from C_n in the Hajian–Kakutani skyscraper

Given a column C_n with height h_n, write $C_n = \{I_{n,0}, \ldots, I_{n,h_n-1}\}$, where $X_n = \bigsqcup_{j=0}^{h_n-1} I_{n,j}$ is an interval. To construct C_{n+1} first cut each level $I_{n,j}$ of C_n into two equal-length subintervals $I_{n,j}^{[0]}$ and $I_{n,j}^{[1]}$. Then choose $2h_n$ new intervals abutting the right endpoint of X_n, each of the same length as any of the intervals $I_{n,j}^{[i]}$, and denote these spacers by $K_{1,0}, \ldots, K_{1,2h_n-1}$ (we will see later why we choose the first subscript to be 1). The levels of C_{n+1} are then

$$I_{n+1,0} = I_{n,0}^{[0]}, \ldots, I_{n+1,h_n-1} = I_{n,h_n-1}^{[0]},$$
$$I_{n+1,h_n} = I_{n,0}^{[1]}, \ldots, I_{n+1,2h_n-1} = I_{n,h_n-1}^{[1]},$$
$$I_{n+1,2h_n} = K_{1,0}, \ldots, I_{n+1,4h_n-1} = K_{1,2h_n-1}.$$

Then $T_{C_{n+1}}$ is defined as the translation that sends each level in C_{n+1} to the level above it. It is clear that $T_{C_{n+1}}$ agrees with T_{C_n} wherever this map was defined and $T_{C_{n+1}}$ is now defined on $I_{n,2h_n}^{[0]}, I_{n,2h_n}^{[1]}$, and all the new spacer levels except the top one. Write $X_{n+1} = \bigsqcup_{j=0}^{4h_n-1} I_{n+1,j}$; it is left to the reader to verify that X_{n+1} is an interval and $\lambda(X_{n+1}) = 2\lambda(C_n) = 2^{n+1}$. It follows that the transformation $T(x) = \lim_{n \to \infty} T_{C_n}(x)$ is defined on $X = \bigcup_{n=0}^{\infty} X_n = [0, \infty)$.

There is another picture that is useful to keep in mind for this transformation. We think of a "tower" (or column) defined over the unit interval in the following way. Over the subinterval $[1/2, 1)$ place two intervals of length $1/2$. These intervals are chosen to be $[1, 3/2)$ and $[3/2, 2)$. We think of the transformation as moving points up as long as there is an interval to go to. Next, above the interval $[7/4, 2)$ place 8 intervals of length $1/4$.

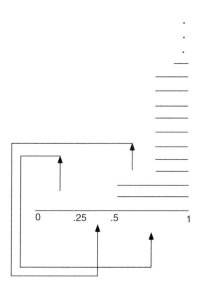

Figure 3.17. The Hajian–Kakutani transformation

A set W is said to be **weakly wandering** under an invertible transformation T if there is a sequence $\{n_i\}_{i=1}^{\infty} \geq 0$ such that
$$T^{n_i}(W) \cap T^{n_j}(W) = \emptyset, \text{ for } n_i \neq n_j.$$
The weakly wandering set W is said to be **exhaustive** if for the weakly wandering sequence $\{n_i\}$
$$X = \bigcup_{i=1}^{\infty} T^{n_i}(W) \bmod \lambda.$$
The reader is asked in the exercises to verify that a finite measure-preserving transformation does not admit weakly wandering sets. A related notion is that of a wandering set. A set W is called **wandering** if $T^{-n}(W) \cap T^{-m}(W) = \emptyset$ for all $m \neq n$, for $n, m \in \mathbb{N}$. Evidently, a wandering set is weakly wandering. In Exercise 6 the reader is asked to show that a recurrent transformation does not admit a wandering set of positive measure. If a transformation is

3.9. Infinite Measure-Preserving Transformations 113

invertible, a wandering set also satisfies: $T^{-n}(W) \cap T^{-m}(W) = \emptyset$ for all $m \neq n$. It can be shown, as an application of the ergodic theorem for infinite measure-preserving transformations, that every ergodic infinite measure-preserving invertible transformation admits a weakly wandering set of positive measure. We prove below the existence of weakly wandering sets for the case of the Hajian–Kakutani transformation.

Theorem 3.9.2. *The Hajian–Kakutani transformation T is infinite measure-preserving, invertible* mod λ, *recurrent and ergodic. Furthermore, T admits an exhaustive weakly wandering set of measure 1.*

Proof. Since the levels in all columns form a sufficient semi-ring, and T sends levels to levels of the same measure, Theorem 3.4.1 implies that T is measure-preserving. It is also clear that T is invertible mod λ. The proof of ergodicity is similar to the one for the dyadic odometer. Let A, B be sets of positive measure. There exists a column C_n and levels I and J in C_n such that I is above J and they are both 1/2-full of A and B, respectively. There is an integer $k > 0$ such that $T^k(J) = I$. Then $\lambda(T^n(A) \cap B) > 0$.

We shall now show that $W = [0, 1)$ is an exhaustive weakly wandering set. Note that the number of spacers that are added to column i to form column $i+1$ is $s_i = 2^{2i+1}$; i.e., there are twice as many as the number of levels of column i. It follows that the sequence of the number of spacers is $2^1, 2^3, 2^5, \ldots$. We shall see that the weakly wandering sequence is an appropriate sum of elements of this sequence.

We construct a sequence of positive integers $\{n_i\}_{i \geq 0}$ such that
$$T^{n_i}(W) \cap T^{n_j}(W) = \emptyset \text{ for } i \neq j.$$
Recall that any integer $i \geq 0$ has a unique base-2 representation of the form
$$i = \delta_0 \cdot 2^0 + \delta_1 \cdot 2^1 + \ldots + \delta_k \cdot 2^k,$$
where $\delta_j \in \{0, 1\}(j = 1, \ldots, k)$ and k is some integer depending on i. Then define
$$n_i = \delta_0 \cdot 2^1 + \delta_1 \cdot 2^3 + \ldots + \delta_k \cdot 2^{2k+1}.$$

The proof now follows by verifying inductively on n that the sets $T^{n_i}(W), i = 1, \ldots, 2^n - 1$, are disjoint, and that $\bigsqcup_{i=1}^{2^n-1} T^{n_i}(W)$ is equal to the union of the levels in C_n. □

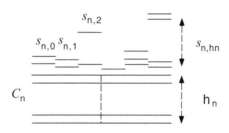

Figure 3.18. Cutting and stacking the C_n column

We now describe a more general type of construction. We again start with C_0 consisting of an interval, usually $[0, 1)$. Given a column C_n, to obtain column C_{n+1}, in addition to the number of cuts r_n, we need to specify the number of spacers that are added on top of each subcolumn. So let $\{s_{n,i}\}$ be a doubly-indexed sequence of nonnegative integers for $n \geq 0$ and $i = 0, \ldots, r_n - 1$. First cut each level of C_n into r_n subintervals to obtain r_n subcolumns of C_n indexed from 0 to $r_n - 1$. Then add on top of the ith-subcolumn $s_{n,i}$ spacers (of the same length as all the levels in that subcolumn). Then extend the transformation to send, by translation, the top level of the ith-subcolumn to the first spacer above, and its top spacer to the bottom level of the $i + 1^{\text{st}}$-subcolumn. The map $T_{C_{n+1}}$ remains undefined on the top level of the last subcolumn if there are no spacers above it, and on the top spacer of the last subcolumn if there are spacers about the last subcolumn. Figure 3.18 illustrates this. In particular we note that the Hajian–Kakutani skyscraper is obtained for $r_n = 2$ and $s_{n,0} = 0, s_{n,1} = 2h_n$. We will call this construction the **cutting and stacking** construction with sequence of **cuts** r_n and sequence of **spacers** $s_{n,i}$. In Chapter 6, we will see constructions where C_0 consists of an interval different from $[0, 1)$.

Exercises

(1) Let T be the Hajian–Kakutani skyscraper transformation. Show that T^2 is not ergodic.

(2) Let T be the Hajian–Kakutani skyscraper transformation. Find an exhaustive weakly wandering set of measure 2 for T. Is there an exhaustive weakly wandering set of infinite measure for T?

(3) Construct an ergodic infinite measure-preserving transformation T such that T^2 is ergodic but T^3 is not ergodic.

(4) Let T be a cutting and stacking construction with $r_n = 2$ and $s_{n,0} = 0, s_{n,1} = 2h_n + 1$. Show that T^n is ergodic for all $n \neq 0$.

(5) Let T be the Hajian–Kakutani skyscraper transformation. Find all integers k so that T^k is ergodic.

(6) Let T be a measure-preserving transformation. Show that T is recurrent if and only if for all measurable sets W such that $\mu(T^{-n}(W) \cap T^{-m}(W)) = 0$ for all $m, n \in \mathbb{N}$ with $m \neq n$, it is the case that $\mu(W) = 0$. (This is very similar to Exercise 3.5.6.)

3.10. Factors and Isomorphism

We study what it means for two dynamical systems to be "the same." The technical term will be isomorphic. The expectation is that two isomorphic dynamical systems will share the same dynamical properties. For example, if a system is ergodic, its isomorphic systems should be ergodic. To start with a simple example, consider a transformation T defined on a space X consisting of two points $X = \{a, b\}$ and a measure defined by $\mu(\{a\}) = \mu(\{b\}) = \frac{1}{2}$ and such that $T(a) = b$ and $T(b) = a$. Define another transformation $S : Y \to Y$ on $Y = \{x, y, z\}$ by $S(x) = y, S(y) = x, S(z) = z$ and a measure ν given by $\nu(\{x\}) = \nu(\{y\}) = \frac{1}{2}, \nu(\{z\}) = 0$. It is clear that $T : X \to X$ and $S : Y \to Y$ should be considered to be isomorphic dynamical systems. First, note that the measure spaces X, μ and Y, ν, after renaming the points and discarding the null sets, are the same. Next note that the map $\phi : X \to Y$ defined by $\phi(a) = x, \phi(b) = y$ preserves the dynamical structure of the maps, in that $T(\phi(a)) = \phi(S(a)), T(\phi(b)) = \phi(S(b))$. Furthermore, the map ϕ is measure-preserving and invertible (after discarding the null set $\{c\}$).

We start with the notion that is used to identify measure spaces. Two measure spaces $(X, \mathcal{S}, \mu)$ and $(X', \mathcal{S}', \mu')$ are said to be **isomorphic** (sometimes we may say measure-theoretically isomorphic or isomorphic mod 0) if there exist measurable sets $X_0 \subset X$ and $X_0' \subset X'$ of **full measure** (i.e., $\mu(X \setminus X_0) = 0$ and $\mu'(X' \setminus X_0') = 0$) and a map

$$\phi : X_0 \to X_0'$$

that is one-to-one and onto and such that

(1) $A \in \mathcal{S}'(X_0')$ if and only if $\phi^{-1}(A) \in \mathcal{S}(X_0)$,
(2) $\mu(\phi^{-1}(A)) = \mu'(A)$ for all $A \in \mathcal{S}'(X_0')$.

We call the map ϕ a **measure-preserving isomorphism mod 0** or **isomorphism mod 0**. Sometimes when we want to be specific about the measures we shall write mod (μ, μ') instead of mod 0.

Example. Let $X = [-1, 1]$, $\mathcal{S} = \mathfrak{L}([-1,1])$, and define a measure μ on $\mathcal{S}$ by $\mu(A) = \frac{1}{2}\lambda(A)$, for $A \in \mathcal{S}$. Clearly, $(X, \mathcal{S}, \mu)$ is a measure space; we claim it is isomorphic to $([0,1], \mathfrak{L}([0,1]), \lambda)$. Indeed, let $\phi : [0,1] \to X$ be given by $\phi(x) = 2x - 1$. Evidently, ϕ is one-to-one and onto and we can compute its inverse $\phi^{-1}(y) = (y+1)/2$. It is not hard to see that ϕ and ϕ^{-1} are measurable. This can be seen by appealing to Exercise 2.3.1 (a result analogous to Theorem 3.4.1 but whose formulation is left to the reader). That ϕ is measure-preserving follows from the fact that it is the composition of a translation (which leaves Lebesgue measure invariant) and a dilation (which expands the measure by 2). So, $\lambda(\phi^{-1}(A)) = \frac{1}{2} \cdot \lambda(A) = \mu(A)$ for all $A \in \mathcal{S}$. (A result analogous to Theorem 3.4.1 also holds here, i.e., for maps from one measure space to another, stating that it suffices to verify the measure-preserving property on a sufficient semi-ring.)

Example. Let F be a closed set in $[0,1]$ of positive Lebesgue measure. Define $\mu(A) = \lambda(A)/\lambda(F)$ for $A \in \mathfrak{L}(F)$. Then $(F, \mathfrak{L}(F), \mu)$ is isomorphic to $([0,1], \mathfrak{L}([0,1]), \lambda)$. The isomorphism we define is an interesting map that has other applications. It is a map $\phi : F \to [0,1]$

3.10. Isomorphism

defined by
$$\phi(x) = \frac{\lambda(F \cap [0, x))}{\lambda(F)}.$$

An important property of ϕ is that it is continuous. Continuity follows from the fact that λ is a nonatomic measure. Let $\varepsilon > 0$. If $|x-y| < \varepsilon$, say $x \geq y$, then

$$\lambda(F) \cdot |\phi(x) - \phi(y)| = \lambda(F \cap [0, x)) - \lambda(F \cap [0, y))$$
$$= \lambda(F \cap [y, x)) \leq |x - y| < \epsilon.$$

This proof shows that ϕ is uniformly continuous. Next we note that ϕ is nondecreasing: if $x < y$, $x, y \in F$, then $\mu(F \cap [0, x)) \leq \mu(F \cap [0, y))$. To see that ϕ is onto, we use the fact that ϕ is nondecreasing. Let $\beta = \sup F$; β is a number in F as $F \subset [0, 1]$ is closed. From the definition of β it is clear that $\phi(\beta) = 1$. Similarly, if $\alpha = \inf F$, then $\alpha \in F$ and $\phi(\alpha) = 0$. Therefore ϕ must be onto. The measure-preserving property can be verified on the sufficient semi-ring of intervals in $[0, 1]$. Let $I = (a, b) \subset [0, 1]$ and choose $a_0, b_0 \in F$ so that $\phi(a_0) = a$, $\phi(b_0) = b$. Then, $\phi^{-1}([0, b)) = \{x \in F : 0 \leq \phi(x) < \phi(b_0)\} = F \cap [0, b_0)$. So

$$\mu(\phi^{-1}([a, b))) = \mu(\phi^{-1}[0, b) \setminus \phi^{-1}[0, a)) = \mu(F \cap [0, b_0) \setminus F \cap [0, a_0))$$
$$= \mu(F \cap [0, b_0)) - \mu(F \cap [0, a_0)) = \phi(b_0) - \phi(a_0)$$
$$= b - a = \lambda([a, b)).$$

It is not hard to verify that properties of a measure space such as sigma-finiteness and the number of atoms, for example, are preserved under isomorphism. The property of being complete (for a measure space), however, is not preserved.

Now we are in a position to define the most important type of measure spaces that we consider. A complete measure space $(X, \mathcal{S}, \mu)$ is called a **Lebesgue space** if it is isomorphic mod 0 to a canonical Lebesgue measure space. We note here that we allow Lebesgue spaces to be of infinite σ-finite measure, while sometimes in the literature they are assumed to be of finite measure. The most interesting measure spaces are Lebesgue spaces and one can make the case that they are the only spaces one needs to be concerned with in ergodic theory. For example, in the exercises the reader is asked to show that the unit interval with Lebesgue measure is isomorphic to the unit square with Lebesgue measure, and that $\mathbb{R}$ is isomorphic to $\mathbb{R}^d$.

A **measure-preserving dynamical system** consists of a Lebesgue measure space $(X, \mathcal{S}, \mu)$ and a measure-preserving transformation $T : X \to X$. In a similar way we define an invertible measure-preserving dynamical system and an invertible measure-preserving dynamical system mod 0.

We now consider the notion of isomorphim for dynamical systems. For the remainder of this section we only consider finite measure-preserving transformations.

Let $(X, \mathcal{S}, \mu, T)$ and $(X', \mathcal{S}', \mu', T')$ be two finite or σ-finite measure-preserving dynamical systems. We say that the two systems are **isomorphic** if there exist measurable sets $X_0 \subset X$ and $X'_0 \subset X'$ of full measure (i.e., $\mu(X \setminus X_0) = 0$ and $\mu'(X' \setminus X'_0) = 0$) with

$$T(X_0) \subset X_0, \ T'(X'_0) \subset X'_0,$$

and there exists a map $\phi : X_0 \to X'_0$, called an **isomorphism**, that is one-to-one and onto and such that for all $A \in \mathcal{S}'(X'_0)$,

(1) $\phi^{-1}(A) \in \mathcal{S}(X_0)$,

(2) $\mu(\phi^{-1}(A)) = \mu'(A)$, and

(3) $\phi(T(x)) = T'(\phi(x))$ for all $x \in X_0$.

The role of the set X_0 is to make precise the fact that the properties of the isomorphism need to hold only on a set of full measure. Property (3) is called **equivariance** and is illustrated the following diagram (note that one typically writes the sets X and X' though the property only holds for a set of full measure in X and X'):

(3.7)
$$\begin{array}{ccc} X & \xrightarrow{T} & X \\ \phi \downarrow & & \downarrow \phi \\ X' & \xrightarrow{T'} & X' \end{array}$$

We shall understand a **dynamical property** to be a property that is invariant under isomorphism. That is, if one dynamical system

3.10. Isomorphism

exhibits the property, then so do all isomorphic systems. (Of course, we refer here to measurable dynamical properties.)

A related and important concept is that of a *factor*. A factor map is similar to an isomorphism except that the factor map is not required to be one-to-one. A dynamical system $(X', \mathcal{S}', \mu', T')$ is a **factor** of $(X, \mathcal{S}, \mu, T)$ if there exist measurable sets $X_0 \subset X$ and $X'_0 \subset X'$ of full measure with

$$T(X_0) \subset X_0, \ T'(X'_0) \subset X'_0,$$

and a map $\phi : X_0 \to X'_0$ that is onto and such that for all $A \in \mathcal{S}'(X'_0)$,

(1) $\phi^{-1}(A) \in \mathcal{S}(X_0)$,

(2) $\mu(\phi^{-1}(A)) = \mu'(A)$ for all $A \in \mathcal{S}'(X'_0)$.

(3) $\phi(T(x)) = T'(\phi(x))$ for all $x \in X_0$.

The main difference with an isomorphism is that the factor map is not required to be one-to-one a.e.

Example. Let $Y = \{p\}$ be a one-point space and let ν be a probability measure on Y. Let $S : Y \to Y$ be the identity transformation. Then $(Y, \mathcal{P}(Y), \nu, S)$ is a factor of any dynamical system $(X, \mathcal{S}, \mu, T)$; it is called the trivial factor. Evidently, T is also a factor of itself.

Example. Let T be the dyadic odometer. From the definition of T it follows that for each $n > 1$ and $i \in \{0, \ldots, h_n - 1\}$, $T(I_{n,i}) = I_{n,i+1}$. This suggests that T acts on the elements of C_n as a rotation. We make this more explicit by defining a map $\phi : X \to Z_n$ by $\phi(x) = i$ if $x \in \bar{I}_{n,i}$ (recall $Z_n = \{0, \ldots, h_n - 1\}$). We claim that ϕ is a factor map and, in this way, see the rotation on n points as a factor of T.

Lemma 3.10.1. *Let S be a factor of T. Then if T is ergodic so is S.*

Proof. Let $\phi : (X, \mu) \to (Y, \nu)$ be the factor map. If A is a strictly invariant set for S, then we claim that $\phi^{-1}(A)$ is a strictly invariant set for T. In fact, $T^{-1}(\phi^{-1}(A)) = \phi^{-1}(S^{-1}(A)) = \phi^{-1}(A)$. Also, $\mu(\phi^{-1}(A)) = \nu(A)$. So $\nu(A) = 0$ or $\nu(A^c) = 0$. □

Exercises

(1) Show that if $(X, \mathcal{S}, \mu, T)$ is a Lebesgue measure-preserving dynamical system, then for any $X_0 \in \mathcal{S}(X)$ with $T^{-1}(X_0) = X_0$, the system $(X_0, \mathcal{S}(X_0), \mu, T)$ is a measure-preserving dynamical system.

(2) Let $(X, \mathcal{S}, \mu)$ be a Lebesgue measure space and let $X_0 \in \mathcal{S}(X)$ with $\mu(X \setminus X_0) = 0$. Suppose there exists a transformation T_0 so that $(X_0, \mathcal{S}(X_0), \mu, T_0)$ is a measure-preserving dynamical system. Show that there exists a transformation $T : X \to X$ so that $T(x) = T_0(x)$ for $x \in X_0$ and $(X, \mathcal{S}, \mu, T)$ is a measure-preserving dynamical system. (T is not unique but differs from T_0 on only a null set.) Show that T_0 is isomorphic to T.

(3) Show that if T and S are isomorphic, then so are T^n and S^n for all $n > 0$.

(4) Show that the doubling map is a factor of the baker's transformation.

(5) Is the dyadic odometer is isomorphic to an irrational rotation?

(6) Let T be an ergodic finite measure-preserving transformation. Show that if T^i is ergodic for $i = 1, \ldots, k-1$ but T^k is not ergodic, then there exists a measurable set A such that the set $T^i(A), i = 1, \ldots, k-1$, are disjoint mod μ and $\bigcup_{i=0}^{k-1} T^i(A) = X$ mod μ.

* (7) Let T be an ergodic finite measure-preserving transformation. Show that T is totally ergodic if and only if it has no factor that is a rotation on n points for any $n > 1$.

3.11. The Induced Transformation

This section treats induced transformations, a useful construction introduced by Kakutani in 1941. While we state the definition in the general case, for most of the section we restrict ourselves to invertible transformations on finite measure spaces, as these illustrate the main ideas and are significantly simpler.

3.11. The Induced Transformation

Let $(X, \mathcal{S}, \mu)$ be a σ-finite measure space and let $T : X \to X$ be a recurrent measure-preserving transformation. Then, for every measurable set A of positive measure there is a null set $N \subset A$ such that for all $x \in A \setminus N$ there is an integer $n = n(x) > 0$ with $T^n(x) \in A$. We call the smallest such n the **first return time** to A, defined by

$$n_A(x) = \min\{n > 0 : T^n(x) \in A\}.$$

We think of $n_A(x)$ as the first "time" that x comes back to A under iteration by T. Since T is recurrent, $n_A(x)$ is defined a.e. for all sets A of positive measure.

As $n_A(x)$ is defined a.e. for all sets A of positive measure, it is possible to define the **induced transformation** of T on A. The induced transformation on a set A of positive measure is denoted by T_A and is defined by

$$T_A(x) = T^{n_A(x)}(x) \text{ for a.e. } x \in A.$$

To define T_A on every point of A one can just let $T_A(x)$ have any fixed value for all x in the null set where n_A is not defined.

Proposition 3.11.1. *Let $(X, \mathcal{S}, \mu)$ be a finite measure space and $T : X \to X$ an invertible measure-preserving transformation. If T is a recurrent transformation and A a set of positive measure, then the induced transformation T_A is measure-preserving on A.*

Proof. For each integer $i \geq 1$ let

$$A_i = \{x \in A : n_A(x) = i\}.$$

The sets A_i are disjoint and

$$A = \bigsqcup_{i=1}^{\infty} A_i \mod \mu.$$

Let $B \subset A$ be measurable. Then,

$$\mu(T_A(B)) = \mu(T_A(\bigsqcup_{n=1}^{\infty} B \cap A_i)) = \mu(\bigsqcup_{i=1}^{\infty} T_A(B \cap A_i))$$

$$= \sum_{i=1}^{\infty} \mu(T^i(B \cap A_i)) = \sum_{i=1}^{\infty} \mu(B \cap A_i)$$

$$= \mu(\bigsqcup_{i=1}^{\infty}(B \cap A_i)) = \mu(B).$$

□

When T_A is defined on a set of finite positive measure A, one usually defines a normalized probability measure μ_A on A by $\mu_A(B) = \mu(B)/\mu(A)$, and T_A is considered with the measure μ_A.

Let T be an invertible, recurrent, measure-preserving transformation. If T is ergodic and A is any set of positive measure, then $\bigcup_{i=1}^{\infty} T^i(A) = X$ mod μ. As before let $A_i = \{x \in A : n_A(x) = i,$ for $i \geq 1\}$. Then $A = \bigsqcup_{i=1}^{\infty} A_i$ mod μ. Therefore, up to a set of μ-measure 0, the set X has the structure given by Figure 3.19. We can observe that if a point x is in A_i, the sets $T(A_i), \ldots, T^{i-1}(A_i)$ are disjoint and outside A. We think of this as a column above i. Once a point reaches the top of the column, its next iterate brings it down to A. Using this, one can obtain a simple proof of the following lemma.

Lemma 3.11.2. *Let $(X, \mathcal{S}, \mu, T)$ be an invertible, finite measure-preserving dynamical system. If T is an ergodic transformation and A a set of positive measure, then the transformation T_A is ergodic on A.*

Proof. Let E, F be sets of positive measure in A. As T is ergodic (and recurrent) there exists an integer $n > 0$ such that $\mu(T^n(E) \cap F) > 0$. So there is a point $x \in E$ such that $T^n(x) \in F$. Let $n_1 = n_A(x), n_2 = n_A(T^{n_1}(x)), \ldots$ until n_k (is the first integer) so that $T^{n_k}(x) \in F$. Then $T_A^k(x) = T^n(x) \in F$, showing that $T_A^k(F) \cap E \neq \emptyset$. Therefore T_A is ergodic. □

3.12. Symbolic Spaces

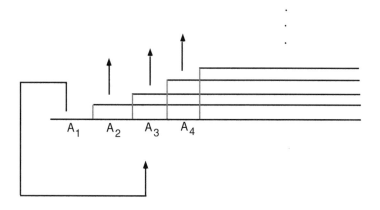

Figure 3.19. An ergodic transformation over a set A

Exercises

(1) Show that every set of positive Lebesgue measure in $\mathbb{R}$ contains a non-Lebesgue measurable subset.

(2) Let T be the dyadic odometer and $A = [0, \frac{1}{2})$. Show that T is isomorphic to the induced transformation T_A on A with normalized measure $\mu_A(B) = \mu(B)/\mu(A)$.

(3) Let $(X, \mathcal{S}, \mu, T)$ be an invertible, recurrent, finite measure-preserving dynamical system. If A is a set of positive measure such that the transformation T_A is ergodic on A, then show that T is ergodic.

3.12. Symbolic Spaces

We study some important symbolic spaces, which at the same time provide interesting examples of metric spaces. Let $N > 1$ be an integer. We let Σ_N^+ denote the set consisting of all infinite sequences of symbols from $\{0, \ldots, N-1\}$. Each element x of Σ_N^+ has the form $x = x(0)x(1) \cdots$ where $x(i) \in \{0, \ldots, N-1\}$ for $i \geq 0$. Sometimes it helps to think of x as a function $x : \mathbb{N}_0 \to \{0, \ldots, N-1\}$, where we write $x(i)$ for the value of x at $i \in \mathbb{N}_0$. We can also view Σ_N^+ as a countably infinite Cartesian product of the finite set $\{0, \ldots, N-1\}$, but we do not emphasize this definition.

While we will concentrate on the space of one-sided infinite sequences Σ_N^+, the space of two-sided infinite sequences, denoted by Σ_N, is also very important. We define the space Σ_N of two-sided infinite sequences of symbols from $\{0, \ldots, N-1\}$ in a similar way: each element of Σ_N can be seen as a function $x : \mathbb{Z} \to \{0, \ldots, N-1\}$, and we write the element x as

$$\cdots x(-2)x(-1)x(0)x(1)x(2) \cdots.$$

The properties of this space are similar to those of Σ_N^+ and are left to the reader as exercises.

We describe a metric in Σ_N^+. Given $x, y \in \Sigma_N^+$, let

$$I(x,y) = \min\{i \geq 0 : x(i) \neq y(i)\}.$$

We think of $I(x,y)$ as the first time where x and y differ. Then define a metric d on Σ_N^+ by

$$d(x,y) = 2^{-I(x,y)},$$

if $x \neq y$ and $d(x,x) = 0$.

Example. Let $x = 0101\overline{01}\cdots$, $y = 0101\overline{11}\cdots$, $z = 1010\overline{11}\cdots$, where $\overline{a}$ means that the pattern a is repeated. Then

$$d(x,y) = 2^{-4}, d(x,z) = 2^0 = d(y,z).$$

The ball $B(x,1)$ consists of all elements of Σ_N^+ starting with 0. Note also that $B(x,1) = B(x, 3/4)$ and $B(x, 3/2) = \Sigma_N^+$.

Lemma 3.12.1. (Σ_N^+, d) *is a compact metric space.*

Proof. It is clear that $d(x,y) = 0$ if and only if $x = y$, and $d(x,y) = d(y,x)$. To show the triangle inequality, note that $I(x,y)$ is the first integer where x and y differ (assuming $x \neq y$). Let $I(x,z) = i$ and $I(z,y) = j$. If $j \geq i$, then y and z cannot differ before i, or $I(x,y) \geq i$. Thus, $I(x,y) \geq \min\{I(x,z), I(y,z)\}$. If $i > j$, then $I(x,y) \geq j$. So in this case also, $I(x,y) \geq \min\{I(x,z), I(y,z)\}$. It follows that

$$d(x,y) \leq \max\{d(x,z), d(z,y)\}.$$

This inequality implies the triangle inequality. So d is a metric. (A metric satisfying this stronger inequality is called an *ultrametric*.)

3.12. Symbolic Spaces

To show compactness, let $\{x_n\}$ be a sequence of elements of Σ_N^+. The idea of the argument can be described in a simpler way by assuming $N = 2$. In the sequence $\{x_n(0)\}$ of first entries there must be infinitely many n so that $x_n(0)$ consists of 0 or of 1 (or both). Suppose 0 is the symbol that appears infinitely often. Let x_{n_1} be the first element of the sequence whose first entry is 0. Now consider the infinite subsequence of those elements that start with 0. Of these, infinitely many must start with 00 or with 01. Say it is 01. Let x_{n_2} be the first element of this new subsequence starting with 01. Of the infinitely many elements starting with 01, there must be infinitely many starting with 010 or 011. In this way we choose the next element of the sequence. Continue in this way to construct the sequence x_{n_i}. Let x be the element of Σ_2^+ such that $x(i) = x_{n_i}(i)$. It is clear that x_{n_i} converges to x in Σ_2^+. The proof for Σ_N^+ is analogous. □

We observe that a metric space that is compact must be complete, as any Cauchy sequence must have a convergent subsequence to a point x of the space. But then one can show that the whole Cauchy sequence must converge to x. The following lemma further explores the topology of Σ_N^+.

Lemma 3.12.2. *Let $B(x, \varepsilon)$ be an open ball in Σ_N^+. Then $B(x, \varepsilon)$ is a closed set.*

Proof. Let $k > 0$ be such that $1/2^k < \varepsilon \leq 1/2^{k-1}$. Then $y \in B(x, \varepsilon)$ if and only if x and y agree at least in the entries $0, \ldots, k$. If x_n is a sequence in $B(x, \varepsilon)$ converging to z, then z must also agree with x in the entries $0, \ldots, k$, so it must be an element of $B(x, \varepsilon)$. □

We present a useful description of balls. Let $w = a_0 \cdots a_{k-1}$ be a word consisting of $k > 0$ symbols from $\{0, \ldots, N-1\}$. The **cylinder** based at w is

$$[a_0 \cdots a_{k-1}] = \{x \in \Sigma_N^+ : x_0 = a_0, \ldots, x_{k-1} = a_{k-1}\}.$$

So, for example, $\Sigma_2^+ = [0] \sqcup [1]$, and $[1] = [10] \sqcup [11]$.

Lemma 3.12.3 gives a useful characterization of the notion of continuity in metric spaces.

Lemma 3.12.3. Let (X,d) and (Y,q) be two metric spaces and $\phi : X \to Y$. The map ϕ is continuous on X if and only if for all $y \in Y$, $\varepsilon > 0$, the set $\phi^{-1}(B(y,\varepsilon))$ is open in X.

Proof. Suppose that ϕ is continuous and let $y \in Y$, $\varepsilon > 0$. If $\phi^{-1}(B(y,\varepsilon)) = \emptyset$, then it is open. Otherwise let z be a point in $\phi^{-1}(B(y,\varepsilon))$; so $q(\phi(z),y) < \varepsilon$. Since ϕ is continuous, as $\varepsilon - q(\phi(z),y) > 0$, there exists $\delta > 0$ so that if $d(z,x) < \delta$, then $q(\phi(z),\phi(x)) < \varepsilon - q(\phi(z),y)$. So if $x \in B(z,\delta)$, then $q(\phi(x),\phi(z)) < \varepsilon - q(\phi(z),y)$. Thus, $q(\phi(x),y) \leq q(\phi(x),\phi(z)) + q(\phi(z),y) < \varepsilon - q(\phi(z),y) + q(\phi(z),y) = \varepsilon$. This means that the open ball $B(x,\delta)$ is contained in the set $\phi^{-1}(B(y,\varepsilon))$, so $\phi^{-1}(B(y,\varepsilon))$ is open.

To show the converse, let $\varepsilon > 0$. Then for every $x \in X$ the set $\phi^{-1}(B(\phi(x),\varepsilon))$ is open. As $x \in \phi^{-1}(B(\phi(x),\varepsilon))$, there exists a number $\delta > 0$ so that the ball $B(x,\delta)$ is contained in $\phi^{-1}(B(\phi(x),\varepsilon))$. This means that whenever z is in $B(x,\delta)$, then $\phi(z)$ is in $B(\phi(x),\varepsilon)$. This implies that ϕ is continuous, completing the proof. □

On any metric space (X,d) one can define the Borel sets in X, denoted $\mathcal{B}(X)$, as the σ-algebra generated by the open sets of X. Given two metric spaces (X,d) and (Y,q) we define a **Borel measurable** transformation $T : X \to Y$ to be a transformation such that the T-inverse image of any Borel set in Y is a Borel set in X. Then we have the following lemma.

Lemma 3.12.4. Let (X,d) and (Y,q) be two metric spaces and $T : X \to Y$ a transformation. If T is continuous, then it is Borel measurable.

Proof. Let
$$\mathcal{A} = \{A \subset Y : T^{-1}(A) \in \mathcal{B}(X)\}.$$
Clearly $\mathcal{A}$ contains the open sets of Y. Show that it is a monotone class and conclude that it contains $\mathcal{B}(Y)$ (Exercise 1). □

Exercises

(1) Complete the proof of Lemma 3.12.4.

(2) Let Ω be the subset of Σ_3^+ consisting of all sequences x that do not have the word 010 at any place. Show that Ω is a closed subset of Σ_3^+.

(3) For $x, y \in \Sigma_N$ define $I(x,y) = \min\{|i| : x(i) \neq y(i)\}$. Define d on Σ_N by $d(x,y) = 2^{-I(x,y)}$, if $x \neq y$ and $d(x,x) = 0$. Show that d is a metric and Σ_N with this metric is compact for $N \geq 2$.

3.13. Symbolic Systems

We study two important transformations defined on symbolic spaces. The first is called the **shift**. Define $\sigma : \Sigma_N^+ \to \Sigma_N^+$ by shifting the entries of $x \in \Sigma_N^+$ to the left. More precisely, the i entry of $\sigma(x)$ is the $i+1$ entry of x:

$$(\sigma(x))(i) = x(i+1).$$

One can easily construct points whose positive orbit under σ is dense. In fact, let w be such that for each integer $k > 0$ it contains all words of length k. So, for example,

$$w = 01000110110000010110011\cdots,$$
$$\sigma^5(w) = 11011000001010011\cdots.$$

It is clear that the positive orbit of w is dense, so σ is topologically transitive. However, it is not minimal as it has many periodic points. For example, $x = 10\overline{10}\cdots$ is a point of period 2.

The next map that we study is the **odometer map** $\tau : \Sigma_N^+ \to \Sigma_N^+$. This is defined by addition by $1 \pmod{N}$ to the first coordinate with a carry to the right. So, for example, if $x = 012\overline{012}\cdots$ in Σ_3^+, then $\tau(x) = 112\overline{012}\cdots, \tau^2(x) = 212\overline{012}\cdots, \tau^3(x) = 022\overline{012}\cdots$.

We now define a probability measure on Σ_N^+. We present the construction in the case when $N = 2$ as this contains all the ideas. The case for arbitrary N is left as an exercise. This measure will be defined on the Borel σ-algebra of Σ_N^+.

First, we define a Borel measurable injective transformation between a Borel subset of $[0, 1)$ and Σ_2^+. As discussed in Section 2.2,

every number x in $[0,1] \setminus D$ has a unique representation in binary form as
$$x = \sum_{i=1}^{\infty} \frac{x_i}{2^i},$$
where $x_i \in \{0,1\}$ and D is the set of binary rational numbers. Let $I_0 = [0,1] \setminus D$. For $x = \sum_{i=1}^{\infty} \frac{x_i}{2^i} \in I_0$ define the map $\psi : I_0 \to \Sigma_2^+$ by
(3.8) $$\psi(x) = x_1 x_2 \cdots x_n \cdots.$$

Then ψ is a one-to-one and continuous map (Exercise 2). It follows that it is Borel measurable and can be used to define a measure ν on Σ_2^+ by
$$\nu(A) = \lambda(\psi^{-1}(A)),$$
for every A in $\mathcal{B}(\Sigma_2^+)$.

Lemma 3.13.1. *The set function ν is a probability measure defined on the Borel σ-algebra of Σ_2^+. Furthermore, for any cylinder $[a_0 \cdots a_{k-1}]$,*
$$\nu([a_0 \cdots a_{k-1}]) = \frac{1}{2^k}.$$

Proof. If $\{A_n\}$ is a disjoint collection of Borel sets in Σ_2^+, then $\{\psi^{-1}(A_n)\}$ are disjoint and Borel in I_0 and furthermore,
$$\nu(\bigsqcup_{n=1}^{\infty} A_n) = \lambda \circ \psi^{-1}(\bigsqcup_{n=1}^{\infty} A_n)$$
$$= \lambda(\bigsqcup_{n=1}^{\infty} \psi^{-1}(A_n))$$
$$= \sum_{n=1}^{\infty} \lambda(\psi^{-1}(A_n)) = \sum_{n=1}^{\infty} \nu(A_n).$$
Also, $\nu(\Sigma_2^+) = \lambda(\psi^{-1}(\Sigma_2^+)) = \lambda(I_0) = 1$. □

Theorem 3.13.2. *Let (X,d) and (Y,q) be two metric spaces and $S : X \to Y$ a map. Let μ be a σ-finite Borel measure on $\mathcal{B}(X)$ and let ν be a σ-finite Borel measure on $\mathcal{B}(Y)$. If for every open set G in Y, $S^{-1}(G)$ is a Borel set in X and*
$$\mu(S^{-1}(G)) = \nu(G),$$
then S is Borel measurable and measure-preserving.

3.13. Symbolic Systems

Proof. The proof is similar to the second proof of Theorem 3.4.1. Write $Y = \bigsqcup Y_n$, where the Y_n are Borel and of finite ν-measure. Set
$$\mathcal{A}_n = \{A : A \in \mathcal{S}(Y_n) \text{ and } T^{-1}(A) \in \mathcal{S}(X), \mu(T^{-1}(A)) = \nu(A)\}.$$
Showing that $\mathcal{A}_n$ contains the Borel sets in Y_n, for each $n \geq 1$, is left to the reader. $\square$

We prove the following lemma for $N = 2$, with the general case left as an exercise.

Corollary 3.13.3. *The transformations σ and τ are both measure-preserving with respect to the Borel probability measure ν on Σ_2^+.*

Proof. We first describe the value of the measure ν on cylinders. By definition, $\nu([0]) = \lambda(\psi^{-1}([0])) = \lambda([0, 1/2)) = 1/2$. In this way we see that
$$\nu([a_0 \cdots a_{k-1}]) = \lambda(\psi^{-1}([a_0 \cdots a_{k-1}]))$$
$$= \lambda([\sum_{i=1}^{k} \frac{a_{i-1}}{2^i}, \sum_{i=1}^{k} \frac{a_{i-1}}{2^i} + \frac{1}{2^k})) = \frac{1}{2^k}.$$
Now,
$$\sigma^{-1}([a_0 \cdots a_{k-1}]) = [a_0 \cdots a_{k-1} 0] \sqcup [a_0 \cdots a_{k-1} 1].$$
So $\nu(\sigma^{-1}([a_0 \cdots a_{k-1}])) = \nu([a_0 \cdots a_{k-1}])$. As open sets in Σ_2^+ are countable unions of cylinders, Theorem 3.13.2 gives that σ is measure-preserving. The proof for τ is similar and left as an exercise. $\square$

Exercises

(1) Show that the set of periodic points for $\sigma : \Sigma_N^+ \to \Sigma_N^+$ is dense.

(2) Show that the map ψ in (3.8) is one-to-one and continuous.

(3) Show that the odometer map τ is continuous and minimal.

(4) Show that the shift map is continuous. (Hint: If x and y agree in k places, for some integer $k > 0$, then $\sigma(x)$ and $\sigma(y)$ agree in $k - 1$ places.)

(5) Complete the proof of Corollary 3.13.3 by showing that τ is measure-preserving on cylinder sets.

(6) Show that the odometer map τ on Σ_2^+ is isomorphic to the odometer map on $[0,1)$.

(7) Show that the shift map σ on Σ_2^+ is isomorphic to the doubling map.

(8) Show that the odometer map is ergodic without appealing to Exercise 6.

(9) Show that the shift map is ergodic without appealing to Exercise 7.

(10) Formulate and prove a lemma analogous to Lemma 3.13.1 in the case of Σ_N.

(11) A map $T : (X, d) \to (Y, q)$ of to metric spaces is said to be a **homeomorphism** if it is invertible and both T and T^{-1} are continuous. Show that the shift map σ on Σ_N is a homeomorphism.

Open Question B. This is an open question due to Furstenberg. Let $T : [0, 1) \to [0, 1)$ be defined by $T(x) = 2x$ and let $S : [0, 1) \to [0, 1)$ be defined by $S(x) = 3x$. Observe that T and S are measure-preserving transformations with respect to Lebesgue measure. Also, the atomic measure supported at the point $\{0\}$ is also invariant for both T and S. There are uncountably many invariant nonatomic Borel measures that are invariant for T, and others that are invariant for S. The open question is whether there is any nonatomic Borel probability measure μ on $[0, 1)$, other than Lebesgue measure, such that μ is invariant for both T and S. This question remains open but special cases are known. a) Find several measures that are invariant for T alone and for S alone. b) Search the literature to find the special solutions to this question by Rudolph. c) Search the literature to find what is currently known about this question. d) Write a paper outlining the partial solution of Rudolph's and later developments of this solution. e) Write a paper describing the latest developments on this question. f) Solve the problem and tell the author.

Chapter 4

The Lebesgue Integral

This chapter presents an introduction to the basic ideas and theorems of Lebesgue integration, one of the great accomplishments of modern analysis. We start by reviewing the Riemann integral, which we assume is familiar to the reader from calculus courses.

4.1. The Riemann Integral

Our aim in this section is not to develop the theory of Riemann integration in detail, but only to point out its main shortcomings, which we later show are resolved by the Lebesgue integral. We start by recalling the definition of the Riemann integral of a function defined on an interval $[a, b]$.

A (point) **partition** of an interval $[a, b]$ is a finite sequence of points $\mathcal{P} = \{x_0, \ldots, x_n\}$ so that $a = x_0 < x_1 < \cdots < x_n = b$. Define the **mesh** of a partition $\mathcal{P}$ by

$$||\mathcal{P}|| = \max\{x_i - x_{i-1} : i = 1, \ldots, n\}.$$

We think of a partition as becoming finer as its mesh decreases to 0. Let $\mathcal{P}' = \{x_1^*, \ldots, x_n^*\}$ denote a sequence of points in partition $\mathcal{P}$ so that $x_i^* \in [x_{i-1}, x_i]$ for $i = 1, \ldots, n$. Let f be a function defined on an interval $[a,b]$. For a given partition $\mathcal{P}$ and points $\mathcal{P}'$ define the sum

$$S(f,\mathcal{P},\mathcal{P}') = \sum_{i=1}^{n} f(x_i^*)(x_i - x_{i-1}).$$

When f is a nonnegative continuous function on an interval $[a,b]$ we can think of the S as approximating the area under f. The function f is said to be **Riemann integrable** if the limit

$$\lim_{||\mathcal{P}||\to 0} S(f,\mathcal{P},\mathcal{P}')$$

exists. We state precisely what is meant by the existence of this limit: there is a real number L so that for all $\varepsilon > 0$ there exists $\delta > 0$ so that $|L - S(f,\mathcal{P},\mathcal{P}')| < \varepsilon$ whenever $||\mathcal{P}|| < \delta$ and $\mathcal{P}'$ is any sequence of points corresponding to $\mathcal{P}$. In this case, the limit L is said to be the Riemann integral of f.

The reader may recall the following definition from a calculus course. When defining the Riemann integral of a continuous function f on an interval $[a,b]$, one first creates an approximation by partitioning $[a,b]$ into n subintervals, and for each subinterval one constructs a rectangle whose height is given by the value of f on some point of the subinterval. These approximating sums are given by the sums $S(f,\mathcal{P},\mathcal{P}')$.

It can be shown that a Riemann integrable function is bounded (Exercise 1). Now we discuss one of the main problems with the Riemann integral. For this we consider a characteristic function given by $f = \mathbb{I}_{\mathbb{Q} \cap [0,1]}$. Evidently, $f = 0$ a.e. Thus, according to our principle of discarding sets of measure zero, f is the same as the zero function, so it seems reasonable to think that f should be integrable and that its integral should be zero. However, one can readily show that f is not Riemann integrable. In fact, for any partition $\mathcal{P}$ there is a sequence of points $\mathcal{P}'$ so that $S(f,\mathcal{P},\mathcal{P}') = 0$ (just let $\mathcal{P}'$ consist of any sequence of points in the subintervals of $\mathcal{P}$ that are not rational). Similarly, given any partition $\mathcal{P}$ there is a sequence of points $\mathcal{P}''$ so that $S(f,\mathcal{P},\mathcal{P}'') = 1$. This implies that f cannot be Riemann integrable.

We continue with our example by constructing a sequence of bounded functions f_n converging to f so that each f_n is Riemann

4.1. The Riemann Integral

integrable. This is an important example as it shows that Riemann integration does not respect limit operations. As we shall see, the Lebesgue integral respects several limit operations.

To define f_n, let $\{a_i\}_{i=1}^{\infty}$ be a listing of the rational numbers in $[0,1]$. Then define f_n so that $f_n(a_i) = 1$ for $i = 1, \ldots, n$ and $f_n(x) = 0$ for all $x \neq a_i$ for all $i \in \{1, \ldots, n\}$. Then $f_n(x) = f(x)$ for all $x \notin \{a_{n+1}, a_{n+2}, \ldots\}$. Therefore $\lim_{n \to \infty} f_n(x) = f(x)$. Next, we observe that each f_n is Riemann integrable. This follows from the fact that f_n is equal to 0 except at a finite number of points. Indeed, one can observe that for any partition $\mathcal{P}$ it follows that $S(f_n, \mathcal{P}, \mathcal{P}') \leq n||\mathcal{P}||$. Therefore $\lim_{||\mathcal{P}|| \to 0} S(f_n, \mathcal{P}, \mathcal{P}') = 0$, which shows that f_n is Riemann integrable.

There exist other minor problems with the Riemann integral that the reader may remember from calculus. One is that it is not defined for unbounded functions. For example, the Riemann integral of the function $1/\sqrt{x}$ is not defined on the interval $[0, 1]$ (let the function take any value at 0). This was easily remedied with the introduction of "improper integrals"; something that is not necessary in the development of Lebesgue integrals as they are naturally defined for a class of functions that include unbounded functions such as $1/\sqrt{x}$ over $(0, 1)$.

We remark that the Lebesgue integral is indeed a generalization of the Riemann integral in that it can be shown that if a function f is Riemann integrable on an interval $[a, b]$, then it is Lebesgue integrable and its Riemann and Lebesgue integrals agree. We state without proof the following characterization of Riemann integration.

Theorem 4.1.1. *Let f be a function defined on an interval $[a, b]$. Then f is Riemann integrable if and only if f is bounded and the set of points in $[a, b]$ where f is not continuous is a null set.*

Exercises

(1) Show that if f is a Riemann integrable function, then it is bounded.

(2) Show that if f is Riemann integrable, then for any constant a the function af is Riemann integrable.

4.2. Measurable Functions

Before defining the Lebesgue integral we study the class of the measurable functions. We will mainly consider real-valued functions defined on a measure space $(X, \mathcal{S}, \mu)$, which for most of our applications the reader may assume is a canonical Lebesgue measure space. While we basically consider only real-valued functions, it is more useful to develop the theory for functions whose values are in the extended real numbers $R^* = [-\infty, +\infty]$ (refer to Appendix A for the algebra with extended real numbers). The generalization to complex valued functions is relatively simple and treated later.

Recall that $\mathbb{I}_A$ denotes the indicator function of a set A. It is reasonable to expect that any definition of a measurable function should be such that the indicator function $\mathbb{I}_A$ will turn out to be measurable if and only if the set A is measurable. Generalizing this to an arbitrary function f, if f were measurable one would expect that (for all numbers a) the set $\{x : f(x) < a\}$ would be measurable.

A function $f : X \to \mathbb{R}^*$ is said to be a **measurable function** or simply **measurable** if for every number $a \in \mathbb{R}$ the set

$$\{x \in X : f(x) < a\}$$

is measurable. Here, of course, "measurable" means measurable as a subset of X, i.e., an element of $\mathcal{S}(X)$. In the case when $\mathcal{S}(X)$ is the σ-algebra of Lebesgue measurable sets of X, then we shall say that f is a **Lebesgue measurable** function. Similarly, in the case when $\mathcal{S}(X)$ is the σ-algebra of the Borel sets we say that f is a **Borel measurable function**. Since Borel sets are Lebesgue measurable, a Borel measurable function is Lebesgue measurable. When it is clear from the context, sometimes we may simply say measurable instead of Lebesgue measurable. The most important class of measurable functions for us will be those that are Lebesgue measurable.

Note that the set $\{x \in X : f(x) < a\}$ is the same as $f^{-1}[-\infty, a)$. (Exercises reviewing properties of inverse images of sets are at the end of this section.)

Proposition 4.2.1. *Let $(X, \mathcal{S}, \mu)$ be a measure space and let $f : X \to \mathbb{R}^*$ be a function. The following are equivalent.*

4.2. Measurable Functions

(1) f is measurable.
(2) For all $a \in \mathbb{R}$, the set $\{x \in X : f(x) \geq a\}$ is measurable.
(3) For all $a \in \mathbb{R}$, the set $\{x \in X : f(x) > a\}$ is measurable.
(4) For all $a \in \mathbb{R}$, the set $\{x \in X : f(x) \leq a\}$ is measurable.
(5) For all $a \in \mathbb{R}$, the set $\{x \in X : f(x) < a\}$ is measurable.

Proof. Suppose that f is a measurable function. Since

$$\{x \in X : f(x) \geq a\} = \{x \in X : f(x) < a\}^c,$$

it follows that $\{x \in X : f(x) \geq a\}$ is measurable. (Here the complement is understood to be in X.) The proof that (2) implies (1) uses the same equality to obtain that $\{x \in X : f(x) < a\}^c$ is measurable, and hence $\{x \in X : f(x) < a\}$ is measurable.

The equivalence of the remaining statements follows through similar arguments from the following set-theoretic equalities. (The proof of these equalities is left to the reader.)

$$\{x \in X : f(x) \leq a\} = \{x \in X : f(x) > a\}^c,$$

$$\{x \in X : f(x) > a\} = \bigcup_{n=1}^{\infty} \{x \in X : f(x) \geq a + \frac{1}{n}\},$$

$$\{x \in X : f(x) < a\} = \bigcup_{n=1}^{\infty} \{x \in X : f(x) \leq a - \frac{1}{n}\},$$

$$\{x \in X : a < f(x)\} = \bigcup_{n=1}^{\infty} \{x \in X : a < f(x) < a + n\}.$$

$\square$

The proof of the following characterization for real-valued functions is left to the reader.

Lemma 4.2.2. *Let $f : X \to \mathbb{R}$ be a real-valued function. f is measurable if and only if for all $a, b \in \mathbb{R}$, the set $\{x \in X : a < f(x) < b\}$ is measurable.*

We briefly discuss the connection between continuous and measurable functions. Let X be a metric space with metric d and let

$f : X \to \mathbb{R}$ be a real-valued function. Recall that f is continuous on X if and only if for all $a, b \in \mathbb{R}$ the set $f^{-1}(a, b)$ is open in X.

Lemma 4.2.3. *Let X be a metric space with metric d and let $f : X \to \mathbb{R}$ be a real-valued function. If f is continuous on $\mathbb{R}$, then f is Borel measurable, hence Lebesgue measurable*

Proof. This is a consequence of Proposition 4.2.1 and Lemma 3.12.4. □

We prove a theorem that says that the property of being a measurable function does not change under basic arithmetic and limit operations. Note that, for example, $\sup_{n \geq 1} f_n$ always exists as an extended real-valued function. Also, not all algebraic operations are defined with the extended real numbers, so that, for example, $f - g$ is not defined when it takes the form $\infty - \infty$.

Theorem 4.2.4.

(1) *If c is a constant and f and g are measurable functions, then the following functions are measurable:*

$$c, \quad cf, \quad f + g, \quad f - g, \quad f \cdot g, \quad \text{and} \quad \frac{f}{g} \quad \text{provided } g \neq 0.$$

(2) *Let $\{f_n\}_{n \geq 1}$ be a sequence of measurable functions. Then the following functions, if they exist, are measurable:*

$$\sup_{n \geq 1} f_n, \quad \inf_{n \geq 1} f_n,$$

$$\limsup_{n \to \infty} f_n, \quad \liminf_{n \to \infty} f_n, \quad \lim_{n \to \infty} f_n.$$

Proof. It is clear that constant functions are measurable. That cf is measurable is left as an exercise. Next, we note that for any $a \in \mathbb{R}$ and any $x \in X$, $f(x) + g(x) < a$ if and only if there is a rational number r such that $f(x) < r$ and $g(x) < a - r$. Therefore we have the equality

$$\{x : f(x) + g(x) < a\}$$
$$= \bigcup_{r \in \mathbb{Q}} (\{x \in X : f(x) < r\} \cap \{x \in X : g(x) < a - r\}).$$

4.2. Measurable Functions

As $\{x \in X : f(x) < r\}$ and $\{x \in X : g(x) < a - r\}$ are measurable sets, it follows that $f + g$ is measurable. A similar argument yields that $f - g$ is measurable.

To show that the product $f \cdot g$ is measurable, we observe that

$$f \cdot g = \frac{1}{4}(f+g)^2 - \frac{1}{4}(f-g)^2.$$

Thus if we show that for any function f, f^2 is measurable, it will follow that $f \cdot g$ is measurable. Now

$$\{x \in X : f^2(x) < a\}$$
$$= \begin{cases} \emptyset, & \text{if } a \leq 0; \\ \{x \in X : f(x) < \sqrt{a}\} \cap \{x \in X : f(x) > -\sqrt{a}\}, & \text{if } a > 0. \end{cases}$$

Therefore f^2 is measurable whenever f is. The proof that f/g is measurable is left to the reader.

To show that $\sup_{n \geq 1} f_n$ is measurable, observe that

$$\{x \in X : \sup f_n(x) > a\} = \bigcup_n \{x \in X : f_n(x) > a\}.$$

The set on the right-hand side is clearly measurable. As

$$\inf_{n \geq 1} f_n = -\sup_{n \geq 1} -f_n,$$

it follows that $\inf_n f_n$ is also measurable.

The other cases reduce to this by noting that

$$\liminf_{n \to \infty} f_n = \sup_{n \geq 1} \inf_{k \geq n} f_k,$$
$$\limsup_{n \to \infty} f_n = \inf_{n \geq 1} \sup_{k \geq n} f_k.$$

Finally, $\lim_{n \to \infty} f_n(x)$ exists if and only if

$$\liminf_{n \to \infty} f(x) = \limsup_{n \to \infty} f_n(x),$$

so $\lim_{n \to \infty} f_n$ is measurable when it exists. $\square$

We define two functions f and g to be **equal almost everywhere** and write

$$f = g \text{ a.e.}$$

if $\mu\{x : f(x) \neq g(x)\} = 0$.

Question. Let f, g be measurable functions and set $A_\alpha = \{x : f(x) < \alpha\}, B_\alpha = \{x : g(x) < \alpha\}$. Show that $f = g$ a.e. if and only if $A_\alpha = B_\alpha \bmod \mu$ for all $\alpha \in \mathbb{R}$.

Lemma 4.2.5. *Let $(X, \mathcal{S}, \mu)$ be a complete measure space. If f is a measurable function and $g = f$ a.e., then g is measurable.*

Proof. Let $D = \{x : f(x) \neq g(x)\}$. Then $\mu(D) = 0$ and D and D^c are measurable. Next, we note that for any $a \in \mathbb{R}$,

$$\{x : g(x) < a\} = (\{x : g(x) < a\} \cap D^c) \cup (\{x : g(x) < a\} \cap D).$$

Now $\{x : g(x) < a\} \cap D^c = \{x : f(x) < a\} \cap D^c$, and this latter set is measurable. Furthermore, $\mu(\{x : g(x) < a\} \cap D) = 0$ and so this set is also measurable. It follows that $\{x : g(x) < a\}$ is measurable, and so g is measurable. □

Dynamical Notions in Terms of Functions

We conclude this section by showing how several dynamical notions we have studied, which were defined in terms of sets, can be defined in terms of functions. We start with an equivalent characterization of measurable transformations. Evidently, for any set A, $\mathbb{I}_A \circ T = \mathbb{I}_{T^{-1}(A)}$. This implies that $\mathbb{I}_A \circ T$ is a measurable function if and only if $T^{-1}(A)$ is a measurable set. It follows that T is a measurable transformation if and only if for all measurable sets A, the function $\mathbb{I}_A \circ T$ is measurable. This remark and the following lemma, which generalizes it, clarify why we use the inverse image of a set in the definition of a measurable transformation.

Lemma 4.2.6. *Let $(X, \mathcal{S}, \mu)$ be a measure space. $T : X \to X$ is a measurable transformation if and only if for all measurable functions $f : X \to \mathbb{R}$, the function $f \circ T$ is measurable. If $f : X \to \mathbb{R}^*$ is measurable and T is a measurable transformation, then $f \circ T$ is measurable.*

Proof. Let T be a measurable transformation and let f be a measurable function. Let $a \in \mathbb{R}$. The proof uses the fact that the set $\{x : f(T(x)) < a\}$ can be written as $T^{-1}(\{z : f(z) < a\})$. As T is a measurable transformation, $\{x : f(T(x)) < a\}$ is a measurable set;

4.2. Measurable Functions

therefore $f \circ T$ is a measurable function. For the converse, apply the remark above to $f = \mathbb{I}_A$, for any measurable set A. □

We note that a set A is strictly T-invariant if and only if $\mathbb{I}_A = \mathbb{I}_A \circ T$. Indeed, if A is strictly T-invariant, then $x \in A$ if and only if $T(x) \in A$. So $\mathbb{I}_A(x) = 1$ if and only if $\mathbb{I}_A(T(x)) = 1$. So $\mathbb{I}_A = \mathbb{I}_A \circ T$. The converse is similar.

The lemma below is the first example of the general principle in ergodic theory that a property stated in terms of measurable sets has an equivalent formulation in terms of measurable functions, and vice versa.

Lemma 4.2.7. *Let T be a measure-preserving transformation on $(X, \mathcal{S}, \mu)$. T is ergodic if and only if for all measurable functions $f : X \to \mathbb{R}$, whenever $f(x) = f(T(x))$ a.e. then f is constant a.e.*

Proof. Suppose T is ergodic. If f is not constant a.e., then there exists a number α such that both sets $\{x : f(x) < \alpha\}$ and $\{x : f(x) > \alpha\}$ have positive measure. However, both sets are strictly invariant mod μ and disjoint; as T is ergodic, they cannot both have positive measure, thus f must be constant a.e. For the converse, note that a set A is strictly invariant mod μ if and only if $\mathbb{I}_A(Tx) = \mathbb{I}_A(x)$ a.e. □

Exercises

(1) Complete the proof of Theorem 4.2.4 by showing that for any measurable functions f and g such that $g \neq 0$, the function f/g is measurable.

(2) Prove Lemma 4.2.2.

(3) Show that a set A is measurable if and only if the characteristic function of A is measurable.

(4) Show that for any sets A and B,
$$\mathbb{I}_{A \cap B} = \mathbb{I}_A \cdot \mathbb{I}_B,$$
$$\mathbb{I}_{A^c} = 1 - \mathbb{I}_A,$$
$$\mathbb{I}_{A \cup B} = \mathbb{I}_A + \mathbb{I}_B - \mathbb{I}_A \cdot \mathbb{I}_B.$$

(5) Show that for any set X and any function $f : X \to \mathbb{R}$, for any sets $A, B \subset \mathbb{R}$,
$$f^{-1}(A \cup B) = f^{-1}(A) \cup f^{-1}(B),$$
$$f^{-1}(A \cap B) = f^{-1}(A) \cap f^{-1}(B),$$
$$f^{-1}(\mathbb{R} \setminus A) = X \setminus f^{-1}(A).$$
What about the analogous equalities for $f(A \cup B)$, $f(A \cap B), f(X \setminus A)$, for $A, B \subset X$. Are they true?

(6) Show that a function $f : \mathbb{R} \to \mathbb{R}$ is a Lebesgue measurable function if and only if for any open set G the set $f^{-1}(G)$ is Lebesgue measurable.

(7) Show that if $g : \mathbb{R} \to \mathbb{R}$ is Lebesgue measurable and $f : \mathbb{R} \to \mathbb{R}$ is continuous, then $f \circ g$ is Lebesgue measurable.

(8) Show that if $f, g : \mathbb{R} \to \mathbb{R}$ are Lebesgue measurable and g is such that for all null sets N, $g^{-1}(N)$ is measurable, then $f \circ g$ is Lebesgue measurable.

(9) Show that if $f : \mathbb{R} \to \mathbb{R}$ is a Lebesgue measurable function, then for all $a \in \mathbb{R}$ the set $\{x : f(x) = a\}$ is measurable.

(10) Let $(X, \mathcal{S}, \mu)$ be a measure space and let $f_n : X \to \mathbb{R}$ be a sequence of measurable functions. Let $\alpha \in \mathbb{R}$. Show that the set
$$\{x \in X : \liminf_{n \to \infty} f_n(x) > \alpha\}$$
is the same as the set
$$\bigcup_{k=1}^{\infty} \bigcup_{\ell=1}^{\infty} \bigcap_{n=\ell}^{\infty} \{x \in X : f_n(x) - \alpha \geq \frac{1}{k}\}.$$
State and proof a similar statement for the set
$$\{x \in X : \limsup f_n(x) < \beta\}.$$

(11) Show that $A = B \mod \mu$ if and only if $\mathbb{I}_A = \mathbb{I}_B$ a.e.

(12) Show that if A is a measurable set, $A = B \mod \mu$ and the measure space is complete, then B is measurable.

(13) Show that if A, B are measurable sets and $A = B \mod \mu$, then $\mu(A) = \mu(B)$.

4.3. The Lebesgue Integral of Simple Functions

(14) Show that a set A is (strictly) T-invariant mod μ if and only if $\mathbb{I}_A(Tx) = \mathbb{I}_A(x)$ a.e.

4.3. The Lebesgue Integral of Simple Functions

When defining the Lebesgue integral, we partition the range of the function, rather than its domain (as is the case in the Riemann integral). We illustrate this in the case of a function $f : [a, b] \to [0, 1)$ defined on an interval $[a, b]$ and whose range is contained in the interval $[0, 1)$. For each fixed $n > 1$, partition $[0, 1]$ into the subintervals $I_k = [k/2^n, (k+1)/2^n)$, for $k = 0, \ldots, 2^n - 1$. Then the sets $A_k = \{x \in [a, b] : f(x) \in I_k\}$ are measurable under the assumption that f is measurable. A measurable function that approximates f (from below) is obtained by defining

$$s(x) = \sum_{k=0}^{2^n - 1} \frac{k}{2^n} \mathbb{I}_{A_k}(x).$$

Our task now is to define an integral of s so that in the cases where the area under s is defined it agrees with the definition of the new integral. To this end, we first consider integrals of sums of characteristic functions, the subject of this section.

Let $(X, \mathcal{S}, \mu)$ be a measure space. It is clear that if one is to define a reasonable notion of integral, the integral of the characteristic function $\mathbb{I}_{[0,2]}$, say, should be 2, as $\mathbb{I}_{[0,2]}$ is zero everywhere except on $[0, 2]$ and the area of its graph over $[0, 2]$ is 2. Note also that 2 is the measure of the set $[0, 2]$. Thus we define the **Lebesgue integral of the characteristic function** $\mathbb{I}_A$ of a measurable set A by

$$\int \mathbb{I}_A \, d\mu = \mu(A).$$

The next class of functions to consider are functions that are sums of characteristic functions with constant coefficients, the so-called simple functions. A function s is said to be a **simple function** if there exist measurable sets $A_1, \ldots, A_k$ and real constants $a_1, \ldots, a_k$ such that

$$s(x) = \sum_{i=1}^{k} a_i \mathbb{I}_{A_i}(x).$$

Question. Show that a function s is simple if and only if it is measurable and takes on only finitely many values.

Suppose that s is a simple function taking on the distinct values $\alpha_1, \ldots, \alpha_n$. Let
$$E_j = \{x : s(x) = \alpha_j\}.$$
Note that, by the definition of the sets E_j,
$$\bigsqcup_{j=1}^{n} E_j = X.$$
Then one can see that
$$s(x) = \sum_{j=1}^{n} \alpha_j \mathbb{I}_{E_j}(x).$$
This is called the **canonical representation of the simple function** s.

Question. Show that $\sum_{j=1}^{n} \alpha_j \mathbb{I}_{E_j}(x)$ is the canonical representation of a simple function if and only if the sets E_j are disjoint and the numbers α_j are distinct.

Now we will define the Lebesgue integral of nonnegative simple functions. Let s be a nonnegative simple function with canonical representation $s(x) = \sum_{j=1}^{n} \alpha_j \mathbb{I}_{E_j}(x)$. Define the **Lebesgue integral of the simple function** s by
$$\int s \, d\mu = \sum_{j=1}^{n} \alpha_j \mu(E_j).$$
If there is some j such that $\alpha_j = 0$ and $\mu(E_j) = \infty$ we agree to write $\alpha_j \mu(E_j) = 0$; this is the only reasonable convention, as it is saying that the integral of the zero function is independent of the measure of the set where it is defined and is always 0. Note that the canonical representation of a simple function is unique up to reordering, and the definition of the integral is independent of the order in which the canonical representation is written.

The characteristic function $\mathbb{I}_A$ has canonical representation as a simple function given by
$$\mathbb{I}_A(x) = 1 \cdot \mathbb{I}_A + 0 \cdot \mathbb{I}_{A^c},$$

4.3. The Lebesgue Integral of Simple Functions

and the definitions of its integral as a characteristic function and as a simple function agree.

The next lemma is used in the proof of the theorem below; it will eventually yield that the integral of a simple function is independent of its representation. The main content of the lemma is due to the fact that the numbers a_i are not assumed to be distinct.

Lemma 4.3.1. *Let*
$$s = \sum_{i=1}^{n} a_i \mathbb{I}_{A_i},$$
and suppose that $A_i \cap A_j = \emptyset$ for $i \neq j$, and that s is nonnegative. Then
$$\int s \, d\mu = \sum_{i=1}^{n} a_i \mu(A_i).$$

Proof. For each k let
$$E_k = \{x : s(x) = a_k\}.$$
Then E_k is the union of all the sets A_i for i such that $a_i = a_k$, i.e.,
$$E_k = \bigcup_{i : a_i = a_k} A_i.$$
Then
$$a_k \mu(E_k) = \sum_{i : a_i = a_k} a_i \mu(A_i),$$
and therefore
$$\int s \, d\mu = \sum_k a_k \mu(E_k) = \sum_i a_i \mu(A_i).$$
$\square$

Theorem 4.3.2.

(1) *If s_1 and s_2 are simple functions that are nonnegative, and α, β are nonnegative real numbers, then*
$$\int \alpha s_1 + \beta s_2 \, d\mu = \alpha \int s_1 \, d\mu + \beta \int s_2 \, d\mu.$$

(2) *If s_1 and s_2 are nonnegative simple functions and $s_1 \leq s_2$, then*
$$\int s_1 \, d\mu \leq \int s_2 \, d\mu.$$

Proof. (1) Write s_1 and s_2 in canonical form as

$$s_1 = \sum_{i=1}^{k} \alpha_i \mathbb{I}_{E_i} \text{ and } s_2 = \sum_{j=1}^{\ell} \beta_j \mathbb{I}_{F_j}.$$

Note that $\bigcup_{i=1}^{k} E_i = \bigcup_{j=1}^{\ell} F_j = X$. The sets $E_i \cap F_j$ are disjoint for all i,j and their union is $\mathbb{R}$. Rename the $k\ell$ sets $E_i \cap F_j$ by A_i, $i = 1, \ldots, n$ for some $n \geq 1$. Then

$$s_1 = \sum_{i=1}^{n} a_i \mathbb{I}_{A_i} \text{ and } s_2 = \sum_{i=1}^{n} b_i \mathbb{I}_{A_i},$$

for some a_i, b_i (now not necessarily distinct). Then, using Lemma 4.3.1,

$$\int \alpha s_1 + \beta s_2 \, d\mu = \int \alpha \sum_{i=1}^{n} a_i \mathbb{I}_{A_i} + \beta \sum_{i=1}^{n} b_i \mathbb{I}_{A_i} \, d\mu$$

$$= \int \sum_{i=1}^{n} (\alpha a_i + \beta b_i) \mathbb{I}_{A_i} \, d\mu$$

$$= \sum_{i=1}^{n} (\alpha a_i + \beta b_i) \mu(A_i)$$

$$= \alpha \sum_{i=1}^{n} a_i \mu(A_i) + \beta \sum_{i=1}^{n} b_i \mu(A_i)$$

$$= \alpha \int a_i \mu(A_i) \, d\mu + \beta \int b_i \mu(A_i) \, d\mu.$$

Part (2) follows from the fact that if $x \in A_i$, then $s_1(x) = a_i \leq b_i = s_2(x)$. □

The following corollary is a direct consequence of Theorem 4.3.2 and shows that the integral of simple functions is independent of the way they are represented, and, in particular, that the disjointness assumption in Lemma 4.3.1 is not necessary.

Corollary 4.3.3. *Let*

$$s = \sum_{i=1}^{m} a_i \mathbb{I}_{A_i},$$

4.4. The Lebesgue Integral of Nonnegative Functions

and assume that s is nonnegative. Then

$$\int s\, d\mu = \sum_{i=1}^{m} a_i \mu(A_i).$$

Exercises

(1) Show that if s_1 and s_2 are simple nonnegative functions and $s_1 = s_2$ a.e., then $\int s_1\, d\mu = \int s_2\, d\mu$.

(2) Complete the proof of Theorem 4.3.2, part (2).

(3) Show that if s is a nonnegative simple function, then $\int_X s\, d\mu = 0$ if and only if $s = 0$ a.e. in X.

(4) Complete the details in the proof of Corollary 4.3.3.

4.4. The Lebesgue Integral of Nonnegative Functions

We are ready to define the Lebesgue integral of a nonnegative measurable function. It is defined as the supremum of the integrals of all nonnegative simple functions that approximate f from below; namely, the **Lebesgue integral of a nonnegative function** $f : X \to [0, \infty]$ is defined by

$$\int f\, d\mu = \sup\left\{ \int s\, d\mu : s \text{ is simple and } 0 \leq s \leq f \right\}.$$

It is clear that sometimes this integral may attain the value ∞. We say that the function f is **integrable** or **Lebesgue integrable** if $\int f\, d\mu < \infty$. Now we define the integral of a function f over a measurable set A. First note that the function $f \cdot \mathbb{I}_A$ is equal to f on A and equal to 0 elsewhere. Thus it is reasonable to define

$$\int_A f\, d\mu = \int f \cdot \mathbb{I}_A\, d\mu.$$

We may sometimes write $\int_X f\, d\mu$ instead of $\int f\, d\mu$ and, when it might be useful to emphasize the variable x, we write $\int f(x)\, d\mu(x)$ for $\int f\, d\mu$. A nonnegative measurable function f is **integrable** or **Lebesgue integrable over a measurable set** A if $\int_A f\, d\mu < \infty$. For example, the constant function 1 is not integrable over $\mathbb{R}$

(with Lebesgue measure) but it is integrable over any set of finite measure A.

We now prove two of the most important theorems in the theory of Lebesgue integration.

Lemma 4.4.1 (Fatou's Lemma). *Let $\{f_n\}$ be a sequence of nonnegative measurable functions. Then*

$$\int \liminf_{n\to\infty} f_n \, d\mu \leq \liminf_{n\to\infty} \int f_n \, d\mu.$$

Proof. Let $f(x) = \liminf_{n\to\infty} f_n(x)$ for all $x \in X$. We have already seen that f is a measurable nonnegative function. Let s be any nonnegative simple function such that

$$s(x) \leq f(x) \text{ for all } x \in X.$$

Write s in canonical form as

$$s(x) = \sum_{j=1}^{m} \alpha_j \mathbb{I}_{E_j}(x).$$

Then for each $x \in E_j$, $j \in \{1, \ldots, m\}$,

$$\alpha_j \leq f(x).$$

Thus for any real number a, $0 < a < 1$, as $a\alpha_j < f(x)$ it follows that $a\alpha_j \leq f_n(x)$ for all $n > k$, for some integer k. This suggests defining the sets

$$F_{j,k} = F_{j,k}(a) = \{x \in E_j : f_n(x) \geq a\alpha_j, \text{ for all } n > k\}.$$

Evidently, $\bigcup_{k>0} F_{j,k}(a) = E_j$ for all $0 < a < 1$. Now define the simple functions

$$s_{a,k}(x) = \sum_{j=1}^{m} a\alpha_j \mathbb{I}_{F_{j,k}}(x).$$

Then for each $x \in X$, $n > k$,

$$s_{a,k}(x) \leq f_n(x).$$

4.4. The Lebesgue Integral of Nonnegative Functions

So,

$$\int s_{a,k}\, d\mu \leq \int f_n\, d\mu,$$

$$\sum_{j=1}^{m} a\alpha_j \mu(F_{j,k}) \leq \int f_n\, d\mu,$$

$$\sum_{j=1}^{m} a\alpha_j \mu(F_{j,k}) \leq \liminf_{n\to\infty} \int f_n\, d\mu,$$

$$\sum_{j=1}^{m} a\alpha_j \mu(E_j) \leq \liminf_{n\to\infty} \int f_n\, d\mu,$$

where the last inequality is obtained letting $k \to \infty$ and using Proposition 2.5.2. Since this holds for all $a < 1$,

$$\sum_{j=1}^{m} \alpha_j \mu(E_j) \leq \liminf_{n\to\infty} \int f_n\, d\mu,$$

or

$$\int s\, d\mu \leq \liminf_{n\to\infty} \int f_n\, d\mu.$$

Taking the supremum over all simple functions s with $s \leq f$ completes the proof. $\square$

Theorem 4.4.2 (Monotone Convergence Theorem)**.** *Let f_n be a sequence of nonnegative measurable functions such that*

$$0 \leq f_1 \leq f_2 \leq \ldots \leq f_n \leq \ldots.$$

If

$$f(x) = \lim_{n\to\infty} f_n(x) \quad \text{a.e.},$$

then

$$\int f\, d\mu = \lim_{n\to\infty} \int f_n\, d\mu.$$

Proof. Since $f_n \leq f$, then $\int f_n \, d\mu \leq \int f \, d\mu$ (since the set $\{\int s : s \leq f_n\}$ is a subset of $\{\int s : s \leq f\}$). Then

$$\limsup_{n \to \infty} \int f_n \, d\mu \leq \int f = \int \lim_{n \to \infty} f_n \, d\mu$$

$$= \int \liminf_{n \to \infty} f_n \, d\mu \leq \int \liminf_{n \to \infty} f_n \, d\mu$$

$$\leq \liminf_{n \to \infty} \int f_n \, d\mu.$$

□

Corollary 4.4.3. *Let $\{f_n\}$ be a sequence of nonnegative measurable functions such that $\sum_{n=1}^{\infty} f_n(x)$ converges for a.e. x in X. Then*

$$\int \sum_{n=1}^{\infty} f_n(x) \, d\mu(x) = \sum_{n=1}^{\infty} \int f_n(x) \, d\mu(x).$$

Proof. Let $g_k(x) = \sum_{n=1}^{k} f_n(x)$. Then $\{g_k\}$ is a monotone sequence of nonnegative measurable functions converging to $\sum_{n=1}^{\infty} f_n(x)$. By the Monotone Convergence Theorem,

$$\int \sum_{n=1}^{\infty} f_n(x) \, d\mu(x) = \int \lim_{k \to \infty} \sum_{n=1}^{k} f_n(x) \, d\mu(x)$$

$$= \lim_{k \to \infty} \int \sum_{n=1}^{k} f_n(x) \, d\mu(x)$$

$$= \lim_{k \to \infty} \sum_{n=1}^{k} \int f_n(x) \, d\mu(x)$$

$$= \sum_{n=1}^{\infty} \int f_n(x) \, d\mu(x).$$

□

We now prove a useful approximation property.

Lemma 4.4.4. *Let $f : X \to [0, \infty]$ be a nonnegative measurable function. Then, there exists a sequence $\{s_n\}_{n \geq 1}$ of nonnegative simple*

4.4. The Lebesgue Integral of Nonnegative Functions

functions such that for each x,
$$0 \leq s_1(x) \leq s_2(x) \leq \ldots \leq s_n(x) \leq \cdots \leq f(x),$$
$$\lim_{n \to \infty} s_n(x) = f(x).$$

Furthermore, if f is integrable, then the s_n are integrable.

Proof. For each $n \geq 1$ and $k = 1, \ldots, n2^n$ define the sets
$$E_{n,k} = \{x : \frac{k-1}{2^n} \leq f(x) < \frac{k}{2^n}\}.$$
(Note that, for each $n \geq 1$, the intervals $\{[\frac{k-1}{2^n}, \frac{k}{2^n})\}_{k=1}^{n2^n}$ form a partition of $[0, n)$ into subintervals of length $\frac{1}{2^n}$.) It is clear that each of the sets $E_{n,k}$ is measurable. Then set
$$s_n(x) = \sum_{k=1}^{n2^n} \frac{k-1}{2^n} \mathbb{I}_{E_{n,k}}(x) \text{ if } f(x) < \infty, \text{ and } s_n(x) = n \text{ otherwise.}$$

Clearly, each s_n is a simple measurable function. Now let $\varepsilon > 0$ and $x \in X$. If $f(x) < \infty$, choose $n \geq 1$ such that $f(x) < n$ and $\frac{1}{2^n} < \varepsilon$. Thus for some $k \in \{1, \ldots, n2^n\}$, $f(x) \in E_{n,k}$ and $0 < f(x) - s_n(x) < \varepsilon$. If $f(x) = \infty$, $s_n(x) = n$, which converges to $f(x)$ as $n \to \infty$. □

With the aid of Theorem 4.3.2, Theorem 4.4.2 and Lemma 4.4.4 one can prove the following result.

Theorem 4.4.5. *Let f, g be nonnegative integrable functions. Then for any nonnegative constants α, β, $\alpha f + \beta g$ is integrable and*
$$\int \alpha f + \beta g \, d\mu = \alpha \int f \, d\mu + \beta \int g \, d\mu.$$

The following lemma further clarifies why we use the inverse image of a set in the definition of measure-preserving and can be thought of as a change of variables formula.

Lemma 4.4.6. *A measurable transformation T on a σ-finite measure space is measure-preserving if and only if for all integrable nonnegative functions $f : X \to [0, \infty)$,*

(4.1) $$\int f \circ T \, d\mu = \int f \, d\mu.$$

Proof. Let T be a measurable transformation and suppose equation (4.1) is satisfied. For any measurable set A of finite measure write $f = \mathbb{I}_A$, an integrable function. Then $f \circ T = \mathbb{I}_{T^{-1}(A)}$. Since $\int f \circ T \, d\mu = \mu(T^{-1}(A))$ and $\int f \, d\mu = \mu(A)$, it follows that $\mu(T^{-1}(A)) = \mu(A)$. Now, when A has infinite measure, write $A = \bigcup_{n=1}^{\infty} A_n$, where the A_n are pairwise disjoint and of finite measure. Then $\mu(T^{-1}(\bigcup_{n=1}^{\infty} A_n)) = \sum_{n=1}^{\infty} \mu(T^{-1}(A_n)) = \sum_{n=1}^{\infty} \mu(A_n) = \mu(A)$, and therefore T is measure-preserving. For the converse, assume that T is measure-preserving and first note that (4.1) holds when f is the characteristic function of a measurable set A. Now let f be an arbitrary integrable nonnegative function. By Lemma 4.4.4, there is a sequence of simple functions $\{s_n\}_{n \geq 1}$ such that $s_1 \leq \ldots \leq s_n \leq \ldots \leq f$ and $\lim_{n \to \infty} s_n(x) = f(x)$ for all $x \in X$. As the s_n are integrable and are finite sums of characteristic functions, for each $n \geq 1$, $\int s_n \circ T \, d\lambda = \int s_n \, d\lambda$. An application of the Monotone Convergence Theorem (Theorem 4.4.2) shows that $\int f \circ T \, d\mu = \int f \, d\mu$. $\square$

Exercises

(1) Show that if f and g are nonnegative measurable functions such that $f(x) \leq g(x)$ for a.e. x in a measurable set A, then $\int_A f \, d\mu \leq \int_A g \, d\mu$. Furthermore, show that if g is integrable over A, then so is f.

(2) Let f be a nonnegative measurable function and let A be a measurable set. Show that $\int_A f \, d\mu = 0$ if and only if $f = 0$ a.e. on A.

(3) Let f be a nonnegative measurable function and A, B measurable sets. Show that if $A \subset B$, then $\int_A f \leq \int_B f$.

(4) Let f be a nonnegative measurable function. Show that for any sequence of disjoint measurable sets $\{A_j\}$,

$$\int_{\bigsqcup A_j} f \, d\mu = \sum_j \int_{A_j} f \, d\mu.$$

(5) Find a sequence of functions f_n so that the inequality in Fatou's lemma is strict.

4.5. Application: The Gauss Transformation

We start with the following lemma that is used to define a wide class of measures.

Lemma 4.5.1. *Let $(X, \mathcal{S}, \mu)$ be a σ-finite measure space. Let $h : X \to \mathbb{R}$ be a nonnegative measurable function and for any measurable set A, define*
$$\nu(A) = \int_A h\, d\mu.$$
Then ν is a σ-finite measure on the σ-algebra $\mathcal{S}$. Furthermore, if h is integrable, then ν is finite.

Proof. If $\{A_i\}$ is a disjoint collection in $\mathcal{S}$,
$$\nu(\bigsqcup_{i=1}^{\infty} A_i) = \int_{\bigsqcup_{i=1}^{\infty} A_i} h\, d\mu = \sum_{i=1}^{\infty} \int_{A_i} h\, d\nu$$
$$= \sum_{i=1}^{\infty} \int_{A_i} h\, d\nu = \sum_{i=1}^{\infty} \nu A_i.$$

As μ is σ-finite, there exist sets A_i such that h is integrable on each A_i and $X = \bigcup_{i=1}^{\infty} A_i$. So ν is σ-finite. □

There are many naturally arising transformations defined on the real line or the unit interval (such as the Gauss map) which, while not preserving Lebesgue measure, do preserve a measure ν given by the integral of some density with respect to Lebesgue measure. For example, in Exercise 3 the reader is asked to show that an invariant measure for the transformation $T(x) = 4x(1-x)$ on $[0,1]$ is given by $\nu(A) = \int_A h\, d\lambda$, where $h(x) = \frac{1}{2\sqrt{x(1-x)}}$. Such a function h is called an invariant density function for T. We note that while in this example the function h satisfies that $h > 0$ a.e., sometimes it is only the case that h satisfies that $h \geq 0$ a.e.

It is clear that if a measure ν is given by $\nu(A) = \int_A h\, d\mu$ with $h \geq 0$ μ-a.e., then for all measurable sets A, if $\mu(A) = 0$, then $\nu(A) = 0$. If $(X, \mathcal{S}, \mu)$ is a σ-finite measure space and ν is a measure defined on $\mathcal{S}$, we say that ν is **absolutely continuous** with respect to μ if whenever $\mu(A) = 0$, then $\nu(A) = 0$. A theorem of Radon-Nikodym that we do not cover here states that if ν is absolutely continuous with respect to μ, then there exists a measurable function $h \geq 0$ μ-a.e. such that $\nu(A) = \int_A h \, d\mu$.

In general, determining whether a transformation admits an absolutely continuous invariant measure (the absolute continuity of the measure is usually with respect to Lebesgue measure) is a difficult question. While this is a topic not covered in this book, we briefly discuss some of the ideas. The assumption that is made on such transformations is that of being *nonsingular*. A measurable transformation $T : X \to X$ is said to be **nonsingular** if $\mu(A) = 0$ if and only if $\mu(T^{-1}(A)) = 0$. (Some authors consider a weaker condition: T is said to be **negatively nonsingular** with respect to μ if for all $A \in \mathcal{S}$, $\mu(A) = 0$ if $\mu(T^{-1}(A)) = 0$.) Clearly, a measure-preserving transformation is nonsingular. But, for example, the map $T(x) = x^2, x \in [0,1]$, is nonsingular with respect to Lebesgue measure, while it is not measure-preserving for Lebesgue measure. An important question to consider is whether a nonsingular transformation admits an absolutely continuous (or equivalent) finite or σ-finite invariant measure. In this section we study a transformation that admits an equivalent invariant finite measure and has applications to continued fractions. The equivalent invariant measure is due to Gauss.

We now define a transformation on the unit interval that was studied by Gauss and is called the **Gauss map**. For $x \in [0,1)$ define

$$T(x) = \frac{1}{x} - \lfloor \frac{1}{x} \rfloor \text{ if } x \neq 0,$$
$$T(0) = 0.$$

(Here $\lfloor x \rfloor$ is the greatest integer $\leq x$.) Evidently, $1/x - \lfloor 1/x \rfloor$ is a number in $(0,1)$. Note that if $\frac{1}{2} < x < 1$, then $1 < \frac{1}{x} < 2$, so $\lfloor \frac{1}{x} \rfloor = 1$ and $T(x) = \frac{1}{x} - 1$. Proceeding in this way we see that

$$T(x) = \frac{1}{x} - n \text{ for } x \in (\frac{1}{n+1}, \frac{1}{n}).$$

So, over the interval $(\frac{1}{n+1}, \frac{1}{n})$, T is a translation by $-n$ of the hyperbola $\frac{1}{x}$. It follows that for $y \in (0,1)$,

(4.2) $$T^{-1}(\{y\}) = \bigcup_{n=1}^{\infty} \{\frac{1}{n+y}\}.$$

4.5. Application: The Gauss Transformation

Proposition 4.5.2. *Let T be the Gauss map. Then T is measure-preserving with respect to the measure on $[0,1)$ given by*

$$\nu(A) = \frac{1}{\ln 2} \int_A \frac{1}{x+1}\, d\lambda(x).$$

Proof. It suffices to verify the measure-preserving property for all intervals $[a,b)$ in $[0,1)$. Note, however, that $[a,b) = [0,b) \setminus [0,a)$. So it suffices to show that for all $y \in (0,1)$, $\nu(T^{-1}[0,y)) = \nu[0,y)$. Now,

$$\nu[0,y) = \frac{1}{\ln 2} \int_{[0,y)} \frac{1}{1+x}\, d\lambda(x)$$

$$= \frac{1}{\ln 2} \ln(1+x)\big|_0^y$$

$$= \frac{1}{\ln 2} \ln(1+y).$$

From (4.2),

$$T^{-1}([0,y)) = \bigsqcup_{n=1}^{\infty} [\frac{1}{n+y}, \frac{1}{n}) \quad \text{mod } \lambda.$$

Then

$$\nu(T^{-1}[0,y)) = \frac{1}{\ln 2} \int_{\bigsqcup_{n=1}^{\infty}[\frac{1}{n+y},\frac{1}{n})} \frac{1}{1+x}\, d\lambda(x)$$

$$= \frac{1}{\ln 2} \sum_{n=1}^{\infty} \int_{[\frac{1}{n+y},\frac{1}{n})} \frac{1}{1+x}\, d\lambda(x)$$

$$= \frac{1}{\ln 2} \sum_{n=1}^{\infty} \ln\left(\frac{1+1/n}{1+1/(n+y)}\right)$$

$$= \frac{1}{\ln 2} \lim_{n\to\infty} \sum_{i=1}^{n} \ln\frac{i+1}{i} + \ln\frac{i+y}{i+1+y}$$

$$= \frac{1}{\ln 2} \lim_{n\to\infty} \ln \prod_{i=1}^{n} \frac{i+1}{i} \cdot \frac{i+y}{i+1+y}$$

$$= \frac{1}{\ln 2} \ln(1+y).$$

$\square$

We conclude with a brief description of the Gauss map with continued fractions. The reader is probably familiar with the expansion of the number known as the golden mean as

$$\text{(4.3)} \qquad \frac{1+\sqrt{5}}{2} = 1 + \cfrac{1}{1 + \cfrac{1}{1 + \cfrac{1}{1+\cdots}}}.$$

Here the equality is understood in the sense that the fractions on the right-hand side of (4.3) converge to the number $(1+\sqrt{5})/2$. To discuss this in more detail we first note that all the complexity of the question is already present for numbers in $(0,1)$. So let x be a number in $(0,1)$ and let us try to write its expansion such as in (4.3). First write

$$x = \frac{1}{n_1 + x_1},$$

where n_1 is a positive integer. If $x_1 \neq 0$, write

$$x_1 = \frac{1}{n_2 + x_2},$$

where n_2 is a positive integer. If x is irrational, then $x_2 \neq 0$ and one can continue to write

$$x = \cfrac{1}{n_1 + \cfrac{1}{n_2 + \cfrac{1}{n_3 + \cdots}}},$$

where $\{n_i\}$ is a sequence of positive integers. This expression is called the **continued fraction expansion** of x. If x were a rational number this expansion would terminate. When x is an irrational number it can be shown that the coefficients n_i in the expansion are unique. Now note that $n_1 = \lfloor \frac{1}{x} \rfloor$, so $x_1 = T(x)$. Similarly write

$$x_1 = \frac{1}{n_2 + x_2}.$$

4.6. Lebesgue Integrable Functions

Then $\frac{1}{x_1} = n_2 + x_2$, so $n_2 = \lfloor \frac{1}{x_1} \rfloor = \lfloor \frac{1}{T(x)} \rfloor$. In this way one can verify that the sequence n_i satisfies that

$$n_{i+1} = \lfloor \frac{1}{T^i(x)} \rfloor.$$

In closing we mention that it can be shown that the map T is ergodic, in fact mixing (see [**5**, Section 4]).

Exercises

(1) Show that for any nonnegative integrable function f,

$$\int f(x - \frac{1}{x}) \, d\mu(x) = \int f(x) \, d\mu(x).$$

(This gives another proof that the Boole transformation is measure-preserving.)

(2) Define a measure μ on $\mathbb{R}$ by

$$\mu(A) = \int_A \frac{1}{1+x^2} \, d\lambda(x)$$

for all Lebesgue measurable sets A. Let $T : \mathbb{R} \to \mathbb{R}$, a modified Boole transformation, be defined by $T(x) = \frac{1}{2}(x - \frac{1}{x})$. Show that T is a finite measure-preserving transformation with respect to μ.

(3) Define T on $[0, 1]$ by $T(x) = 4x(1-x)$. Show that T preserves the measure ν given by

$$\nu(A) = \int_A \frac{1}{2\sqrt{x(1-x)}} \, d\lambda(x).$$

4.6. Lebesgue Integrable Functions

This section defines the Lebesgue integral for arbitrary measurable functions. If f is any measurable function one can associate with it the following nonnegative measurable functions:

$$f^+(x) = \begin{cases} f(x), & \text{if } f(x) \geq 0; \\ 0, & \text{otherwise,} \end{cases}$$

and
$$f^-(x) = \begin{cases} -f(x), & \text{if } f(x) \leq 0; \\ 0, & \text{otherwise.} \end{cases}$$

The verification that f^+ and f^- are measurable is left to the reader as an exercise; clearly they are nonnegative functions. Furthermore, from their definition it follows that

$$f(x) = f^+(x) - f^-(x) \text{ and } |f(x)| = f^+(x) + f^-(x).$$

We define the **Lebesgue integral** of f by

$$\int f \, d\mu = \int f^+ \, d\mu - \int f^- \, d\mu,$$

provided $\int f^+ \, d\mu < \infty$ and $\int f^- \, d\mu < \infty$; in this case, we say f is **integrable** or **Lebesgue integrable** and write $f \in L^1(X, \mu)$. $L^1(X, \mu)$ denotes the collection of all functions $f : X \to \mathbb{R}$ that are integrable. If $\int f^+ \, d\mu = \infty$ and $\int f^- \, d\mu < \infty$, then we write $\int f = \infty$; if $\int f^+ \, d\mu < \infty$ and $\int f^- \, d\mu = \infty$, then we write $\int f = -\infty$.

Lemma 4.6.1. *Let f be a measurable function. Then f is integrable if and only if $|f|$ is integrable and*

$$\int f \, d\mu \leq \int |f| \, d\mu.$$

Proof. The function f is integrable if and only if f^+ and f^- are integrable. As $|f| = f^+ + f^-$ and since f^+ and f^- are nonnegative, by Theorem 4.4.5, $|f|$ is integrable. Conversely, since $f^+ \leq |f|$ and $f^- \leq |f|$, if $|f|$ is integrable, then by Exercise 1, f is integrable.

Finally,

$$\left| \int f \right| = \left| \int f^+ - \int f^- \right| \leq \left| \int f^+ \right| + \left| \int f^- \right| = \int |f|.$$

□

Lemma 4.6.2. *Let f and g be measurable functions and let α be a real number.*

(1) *If f is integrable, then αf is integrable and $\int \alpha f = \alpha \int f$.*

(2) *If f and g are integrable, then $f + g$ is integrable and*

$$\int f + g = \int f + \int g.$$

4.6. Lebesgue Integrable Functions

(3) If $f \leq g$, then $\int f \leq \int g$.

(4) If $f = g$ a.e., then $\int f = \int g$.

Proof. Parts (1) and (3) are left as an exercise to the reader.

For part (2), we first note that
$$f + g = (f^+ + g^+) - (f^- + g^-).$$
So
$$(f+g)^+ + (f^- + g^-) = (f^+ + g^+) + (f+g)^-.$$
Then
$$\int (f+g)^+ + \int (f^- + g^-) = \int (f^+ + g^+) + \int (f+g)^-,$$
$$\int (f+g)^+ + \int f^- + \int g^- = \int f^+ + \int g^+ + \int (f+g)^-,$$
$$\int (f+g)^+ - \int (f+g)^- = \int f^+ + \int g^+ - \int f^- - \int g^-,$$
$$\int (f+g) = \int f^+ - \int f^- + \int g^+ - \int g^-,$$
$$\int (f+g) = \int f + \int g.$$

For part (4), if $h = f - g$, then $h = 0$ a.e. Write $h = h^+ - h^-$. If s is a simple function with $0 \leq s \leq h^+$, then $s = 0$ a.e. So from the definition of simple functions, $\int s \, d\mu = 0$. Therefore, $\int h^+ \, d\mu = 0$. Similarly, $\int h^- \, d\mu = 0$. Thus, $\int h \, d\mu = 0$, so $\int f \, d\mu = \int g \, d\mu$. □

We end this section with one of the important theorems of Lebesgue.

Theorem 4.6.3 (Lebesgue's Dominated Convergence Theorem). *Let h be an integrable nonnegative function and let $\{f_n\}$ be a sequence of measurable functions such that*
$$\lim_{n \to \infty} f_n(x) = f(x) \quad \text{a.e.}$$
If
$$|f_n| \leq h \quad \text{a.e. for all } n > 0,$$

then f is integrable and

(4.4) $$\lim_{n\to\infty} \int f_n\, d\mu = \int f\, d\mu.$$

Proof. Clearly each f_n is integrable. We first apply Fatou's Lemma to the nonnegative functions $h - f_n$:

$$\int h - f\, d\mu = \int h - \lim_{n\to\infty} f_n\, d\mu = \int \liminf_{n\to\infty}(h - f_n)\, d\mu$$
$$\leq \liminf_{n\to\infty} \int (h - f_n)\, d\mu = \int h + \liminf_{n\to\infty}\left(-\int f_n\, d\mu\right)$$
$$= \int h - \limsup_{n\to\infty} \int f_n\, d\mu.$$

It follows that
$$\int f\, d\mu \geq \limsup_{n\to\infty} \int f_n\, d\mu.$$

For the remaining inequality we again use Fatou's Lemma with the nonnegative functions $h + f_n$ to obtain

$$\int f\, d\mu \leq \liminf_{n\to\infty} \int f_n\, d\mu.$$

□

Remark. Some authors state Lebesgue's Dominated Convergence Theorem with the additional conclusion (under the same hypotheses) that

(4.5) $$\lim_{n\to\infty} \int |f - f_n|\, d\mu = 0.$$

Clearly, (4.5) implies (4.4). For us (4.5) can be obtained as a consequence of (4.4) using Scheffé's Lemma 5.4.3. The reader is asked to give a direct proof (Exercise 10).

Exercises

(1) Show that if f_1 and f_2 are integrable functions and $f_1 = f_2$ a.e., then $\int f_1\, d\mu = \int f_2\, d\mu$.

(2) Show that $|f| = f^+ + f^-$.

(3) Show that f is integrable if and only if $|f|$ is integrable.

(4) Show that if f is an integrable function, then af is integrable for any real number a. (This is in Lemma 4.6.2.)

(5) Show that if $f \leq g$ a.e., then $\int f \leq \int g$. (This is in Lemma 4.6.2.)

(6) Let f be an integrable function. Show that if $\int_A f \, d\mu = 0$ for all measurable sets A, then $f = 0$ a.e.

(7) Let f be a nonnegative integrable function. Suppose that $\{E_p\}_{p>0}$ is a sequence of decreasing $(E_{p+1} \subset E_p)$ measurable sets. Show that if $\lim_{p \to \infty} \mu(E_p) = 0$, then
$$\int_{\bigcap_{p>0} E_p} f \, d\mu = 0.$$

(8) (Chebyshev's Inequality) Show that for any nonnegative measurable function $f : X \to [0, \infty]$, any constant c and $0 < p < \infty$ $(c > 0)$,
$$\mu\{x : f(x) \geq c\} \leq \frac{1}{c^p} \int f^p \, d\mu.$$

(9) Let $f : X \to \mathbb{R}^*$ be a measurable function. Show that if f is integrable, then $|f(x)| < \infty$ a.e.

(10) Give a proof of (4.5).

4.7. The Lebesgue Spaces: L^1, L^2 and L^∞

The L^p spaces studied in this section are some of the most important spaces in analysis. The elements of these spaces are functions, rather than points in a measure space. There are two operations defined on the L^p spaces: addition of functions and multiplication of functions by a scalar. With these operations the spaces have a natural structure as vector spaces, also called linear spaces. The first spaces we study, L^p for $1 \leq p < \infty$, are defined using an integration condition. We first consider real-valued functions, and later discuss the extension to the case of complex-valued functions, which is rather straightforward.

Let $p \geq 1$ be a real number and let $(X, \mathcal{S}, \mu)$ be a σ-finite measure space. For all cases of interest to us, $(X, \mathcal{S}, \mu)$ will be a Lebesgue space. Define $L^p(X, \mu)$ to be the space consisting of all measurable

functions $f : X \to \mathbb{R}$ such that
$$\int |f|^p \, d\mu < \infty.$$
When $p = 1$ this is the space of all integrable functions. We call $L^2(X, \mu)$ the space of **square integrable** functions. We may write L^p for $L^p(X, \mu)$ when the space and measure are clear from the context, and when we need to specify that the functions are complex-valued we write $L^p(X, \mu, \mathbb{C})$.

One can define similar spaces in the case of complex-valued functions. Recall that the field of complex numbers $\mathbb{C}$ consists of numbers of the form $a + bi$ where $a, b \in \mathbb{R}$ and $i^2 = -1$. The **conjugate** of a complex number $c = a + ib$ is defined to be $\bar{c} = a - ib$. The absolute value of a complex number c is $|c| = (a^2 + b^2)^{\frac{1}{2}} = (\bar{c} \cdot c)^{\frac{1}{2}}$. (Note that we use the same notation as for the absolute value of a real number.)

A complex-valued function is a function $f : X \to \mathbb{C}$ of the form $f(x) = f_1(x) + i f_2(x)$, where $f_1, f_2 : X \to \mathbb{R}$ are real-valued functions, called the **real and imaginary part** of f, and written $f_1 = \mathfrak{Re}(f), f_2 = \mathfrak{Im}(f)$, respectively. We say that a complex-valued function f is measurable (integrable) if both f_1 and f_2 are measurable (integrable), and write
$$\int f \, d\mu = \int f_1 \, d\mu + i \int f_2 \, d\mu.$$
Note that if f is complex-valued, its absolute value, $|f|$, is a real-valued function. $L^p(X, \mu, \mathbb{C})$ is defined to be the space of all complex-valued measurable functions $f : X \to \mathbb{C}$ such that
$$\int |f|^p \, d\mu < \infty.$$

Lemma 4.7.1. $L^p(X, \mu, \mathbb{R})$ *is a vector space over $\mathbb{R}$ for each $p \geq 1$ and $L^p(X, \mu, \mathbb{C})$ is a vector space over $\mathbb{C}$ for each $p \geq 1$.*

Proof. We need to verify that L^p is closed under scalar multiplication and addition. The fact that L^p satisfies the axioms of a vector space is straightforward and left to the reader. If $\alpha \in \mathbb{R}$ and $|f|^p$ is integrable, then clearly $|\alpha f|^p$ is integrable. Now let $f, g \in L^p$. Then
$$|f + g|^p \leq (2 \max\{|f|, |g|\})^p \leq 2^p (|f|^p + |g|^p).$$

4.7. The Lebesgue Spaces: L^1, L^2 and L^∞

So $|f+g|^p$ is integrable. □

Now that we know the L^p spaces are vector spaces, we introduce a norm, or size, on each of its elements. For f in L^p define the p-**norm** of f by

$$||f||_p = \left(\int |f|^p \, d\mu\right)^{\frac{1}{p}}.$$

The norm satisfies the properties in Proposition 4.7.2. Among them is the Minkowski inequality, which is an important inequality that does not hold when $0 < p < 1$ and is the main reason why we have assumed $p \geq 1$.

Proposition 4.7.2. *Let $(X, \mathcal{S}, \mu)$ be a σ-finite measure space and let $p \geq 1$. Then*

(1) $||f||_p = 0$ *if and only if* $f = 0$ μ-*a.e.*
(2) $||\alpha f||_p = |\alpha| ||f||_p$ *for all* $\alpha \in \mathbb{R}$ *and* $f \in L^p$.
(3) (*Minkowski Inequality*)

$$||f + g||_p \leq ||f||_p + ||g||_p \text{ for all } f, g \in L^p.$$

Proof. Suppose that $||f||_p = 0$. Then $||f||_p^p = 0$, so $\int |f|^p \, d\mu = 0$. By Exercise 4.4.2, $|f|^p = 0$ a.e., which implies $f = 0$ a.e. The converse is clear.

Now,

$$||\alpha f||_p = \left(\int |\alpha f|^p \, d\mu\right)^{1/p} = |\alpha| \left(\int |f|^p \, d\mu\right)^{1/p} = |\alpha| ||f||_p.$$

For part (3), the case when $p = 1$ follows from the basic triangle inequality for functions. The case when $1 < p < \infty$ uses Hölder's inequality (Theorem 4.7.4) and the proof is left as an exercise. □

A **norm** is a function $||\cdot|| : V \to [0, \infty)$ defined on a (real or complex) vector space V with addition $+$ and scalar multiplication $\cdot$ such that

(1) $||v|| = 0$ if and only if $v = 0$.
(2) $||\alpha \cdot v|| = |\alpha| ||v||$ for all $\alpha \in \mathbb{R}$ (or in $\mathbb{C}$) and $v \in V$.
(3) (Triangle Inequality) $||v + w|| \leq ||v|| + ||w||$ for all $v, w \in V$.

Note that when $||f||_p = 0$, all we can conclude is that $f = 0$ a.e., so that, strictly speaking, $||\cdot||_p$ is not a norm until we identify functions that differ in a null set; this is common practice though sometimes not explicitly stated and in our case it should be clear from the context. A **normed linear space** is a vector space together with a norm defined on it. Thus the L^p spaces, with functions identified when they differ in a null set, are normed linear spaces. (More formally, define a relation between functions so that f is equivalent to g if and only if $f = g$ a.e. One verifies that this is an equivalence relation and lets $\tilde{f}$ be the equivalence class of all the functions equivalent to f. Then the space of all equivalence classes $\tilde{f}$, with $f \in L^p$, is a normed linear space.)

There is another important space that we study in this section. Define $L^\infty = L^\infty(X, \mu)$ to be the space of all measurable functions that are bounded outside a null set. To state this more precisely, we introduce the L^∞ norm $||\cdot||_\infty$. Given a measurable function f, let M_f be the set

$$M_f = \{t : \mu\{x : |f(x)| > t\} = 0\}.$$

One can think of M_f as the set of values that f majorizes only on a set of measure zero. The smallest of such values is defined to be the **essential supremum** of f or its L^∞-norm:

$$||f||_\infty = \inf\{t : t \in M_f\}.$$

Proposition 4.7.3. *$L^\infty(X, \mu)$ with $||\cdot||_\infty$ (under the identification of functions that differ in a null set) is a normed linear space.*

The spaces that will be of interest to us are L^1, L^2 and L^∞. It turns out that L^2 has an additional structure given by an *inner product*. Define the L^2 **inner product** on $L^2(X, \mu, \mathbb{C})$ (or $L^2(X, \mu, \mathbb{R})$) by

$$(f, g) = \int f\bar{g}\, d\mu.$$

The fact that (f, g) exists follows from Hölder's Inequality.

4.7. The Lebesgue Spaces: L^1, L^2 and L^∞

Theorem 4.7.4 (Hölder's Inequality). *Let $(X, \mathcal{S}, \mu)$ be a σ-finite measure space. For each real number $p > 1$ let q be such that*

$$\frac{1}{p} + \frac{1}{q} = 1,$$

and for $p = 1$ let $q = \infty$. For any $p \geq 1$, if $f \in L^p$ and $g \in L^q$, then $f \cdot g$ is integrable and furthermore

$$\int |f \cdot g| \, d\mu \leq \|f\|_p \, \|g\|_q.$$

Proof. First note that when $p = 1$ and $q = \infty$ the inequality follows from the fact that $|f \cdot g| \leq |f| \|g\|_\infty$ a.e. When $p > 1$ we shall use the inequality

$$ab \leq \frac{a^p}{p} + \frac{b^q}{q}$$

for all $\infty > a, b > 0$, $p > 1$ (see Exercise 12). Now let $f \in L^p, g \in L^q$. We may assume that $\|f\|_p \neq 0, \|g\|_q \neq 0$. Apply the inequality with

$$a = \frac{|f(x)|}{\|f\|_p}, b = \frac{|g(x)|}{\|g\|_q}$$

to obtain

$$\frac{|f(x)| \cdot |g(x)|}{\|f\|_p \|g\|_q} \leq \frac{1}{p} \left(\frac{|f(x)|}{\|f\|_p} \right)^p + \frac{1}{q} \left(\frac{|g(x)|}{\|g\|_q} \right)^q.$$

Integrating both sides, we have

$$\int \frac{|f(x)| \cdot |g(x)|}{\|f\|_p \|g\|_q} \, d\mu(x) \leq \frac{1}{p} \int \left(\frac{|f(x)|}{\|f\|_p}\right)^p d\mu(x) + \frac{1}{q} \int \left(\frac{|g(x)|}{\|g\|_q}\right)^q d\mu(x)$$

$$= \frac{1}{p\|f\|_p^p} \int |f(x)|^p \, d\mu(x) + \frac{1}{q\|g\|_q^q} \int |g(x)|^q \, d\mu(x)$$

$$= \frac{1}{p} + \frac{1}{q} = 1.$$

So,

$$\int |f(x)| \cdot |g(x)| \, d\mu(x) \leq \|f\|_p \|g\|_q.$$

□

Note that if $p = 2$, then $q = 2$ satisfies $1/p + 1/q = 1$, so Hölder's inequality implies that if $f, g \in L^2$, then $f \cdot \bar{g}$ is integrable.

Proposition 4.7.5. Let $(X, \mathcal{S}, \mu)$ be a σ-finite measure space. Then for f, g in $L^2(X, \mu, \mathbb{C})$,

(1) $\|f\|_2 = ((f, g))^{1/2}$.
(2) $(\alpha f, g) = \alpha(f, g)$ for all $\alpha \in \mathbb{C}$.
(3) $(f, \alpha g) = \bar{\alpha}(f, g)$ for all $\alpha \in \mathbb{C}$.
(4) $(f, g) = \overline{(g, f)}$.
(5) $(f + g, h) = (f, h) + (g, h)$.
(6) $(f, g + h) = \overline{(f, g)} + \overline{(f, h)}$.

Proof. For part (1),
$$\|f\|_2 = (\int |f|^2 \, d\mu)^{1/2} = (\int f\bar{f} \, d\mu)^{1/2} = ((f, f))^{1/2}.$$
The proofs of the remaining parts are similar and are left to the reader. □

The following lemma is used in the next section. Before we state the lemma, recall from linear algebra that one can use the inner product to define a notion of orthogonality. Given $f, g \in L^2$ we say that f and g are **orthogonal functions** if $(f, g) = 0$. A collection $\mathcal{C}$ of functions in L^2 is said to be an **orthogonal collection** if $(f, g) = 0$ for all $f, g \in \mathcal{C}$, $f \neq g$. A collection $\mathcal{C}$ of functions in L^2 is said to be an **orthonormal collection** if it is an orthogonal collection and $\|f\|_2 = 1$ for all $f \in \mathcal{C}$.

Lemma 4.7.6. Let $(X, \mathcal{S}, \mu)$ be a Lebesgue probability space. If $\mathcal{C}$ is an orthogonal collection of nonzero a.e. functions in $L^2(X, \mu, \mathbb{C})$, then $\mathcal{C}$ is countable.

Proof. We prove the lemma for the case of $L^2([0, 1], \lambda, \mathbb{C})$. Let $\mathcal{D}$ consist of all simple functions that are sums of characteristic functions of dyadic intervals with rational coefficients. Evidently, $\mathcal{D}$ is a countable set. We claim that $\mathcal{D}$ is dense in L^2 in the sense that for any $f \in L^2$ there exists a sequence $s_n \in \mathcal{D}$ so that $\|f - s_n\|_2 \to 0$ as $n \to \infty$. This follows from the fact that finite unions of dyadic intervals approximate measurable sets, and that simple functions approximate measurable functions (see Exercise 14).

4.7. The Lebesgue Spaces: L^1, L^2 and L^∞

Let $\mathcal{C}' = \{h/||h||_2 : h \in \mathcal{C}\}$. Then for $f \neq g \in \mathcal{C}'$,

$$||f - g||_2^2 = \int (f - g) \cdot (\bar{f} - \bar{g}) \, d\lambda$$
$$= \int f \cdot \bar{f} - f \cdot \bar{g} - g \cdot \bar{f} + g \cdot \bar{g} \, d\lambda$$
$$= ||f||_2^2 - (f, g) - (g, f) + ||g||_2^2 = 2.$$

So $||f - g||_2 = \sqrt{2}$. Then for each $f \in \mathcal{C}'$, the collection of open balls $B(f) = \{h \in L^2 : ||h - f|| < \frac{\sqrt{2}}{2}\}$ contains no element of $\mathcal{C}'$ other than f. If $\mathcal{C}'$ were uncountable, $\mathcal{D}$ could not be dense. Therefore $\mathcal{C}$ is countable. □

We end this section with an important theorem, which we state without proof, that implies that the L^p spaces are Banach spaces, a large class of spaces in Functional Analysis defined as complete normed linear spaces. Before the statement, note that as L^p has a norm (again with functions identified a.e.), one can define a metric by $d(f, g) = ||f - g||$, so that L^p becomes a metric space (with functions identified when differing on a null set). As a metric space, L^p is complete if and only if every Cauchy sequence in L^p converges in L^p.

Theorem 4.7.7 (Riesz–Fischer). *Let $(X, \mathcal{S}, \mu)$ be a measure space (a proof can be found in [11, Theorem 13.5]). For each $p \geq 1$, L^p is a complete space.*

Exercises

(1) Show that if $f : X \to \mathbb{C}$ is integrable, then $\bar{f}$ is integrable and $\int f = \overline{\int \bar{f}}$.

(2) Show that if $\mu(X) < \infty$, then $L^p(X, \mu) \subsetneq L^1(X, \mu)$ for all $p > 1$.

(3) Consider the following functions: $f(x) = 1/x$, $g(x) = 1/x^2$, $h(x) = 1/\sqrt{x}$, $j(x) = \sin(x)$. Determine whether each function belongs to the following spaces: $L^1([0,1])$, $L^1((1,\infty))$, $L^2([0,1])$, $L^2([1,\infty))$, $L^1(\mathbb{R})$, $L^2(\mathbb{R})$, $L^\infty(\mathbb{R})$.

(4) Show that the functions $\phi_k : [0,1] \to \mathbb{C}$, $k \in \mathbb{Z}$, defined by $\phi_k(t) = \exp(2\pi k i t)$, are in $L^2([0,1], \lambda, \mathbb{C})$ and form an orthonormal collection.

(5) Complete the proof of Minkowski's Inequality in Proposition 4.7.2.

(6) Prove Proposition 4.7.3.

(7) Show that a function $f : X \to \mathbb{C}$ is integrable if and only if $|f|$ is integrable.

(8) Complete the proof of Proposition 4.7.5.

(9) Give a sequence of measurable functions $\{f_n\}$ so that $||f_n|| \to 0$ as $n \to \infty$ (in this case we say that f_n converges to 0 in $||\cdot||_1$ norm) but $\lim_{n\to\infty} f_n \neq 0$ a.e.

(10) Let f be an integrable function in L^p and $p \geq 1$. Show that for any $\varepsilon > 0$ there exists a bounded function g in L^p so that $||f - g||_p < \varepsilon$.

(11) For any $f, g \in L^1$ define $f \sim g$ if $\mu\{x : f(x) \neq g(x)\} = 0$. Show that $\sim$ is an equivalence relation.

(12) Prove the following inequality, called Young's Inequality:
$$ab \leq \frac{a^p}{p} + \frac{b^q}{q}$$
for all $a, b \geq 0$, $p, q > 1$, such that $1/p + 1/q = 1$.

(13) Say that a σ-finite measure space $(X, \mathcal{S}, \mu)$ has a **countable basis** if there is a countable family of measurable sets $\mathcal{C}$ so that for every $A \in \mathcal{S}$, $\mu(A) < \infty$ and $\varepsilon > 0$, there exists $C \in \mathcal{C}$ so that $\mu(A \triangle C) < \varepsilon$. Show that a Lebesgue measure space has a countable basis.

(14) Let $(X, \mathcal{S}, \mu)$ be a probability measure space with a countable basis. Show that $L^2(X, \mathcal{S}, \mu)$ is separable under the L^2 metric in the sense that there exists a countable family $\mathcal{F}$ of L^2 functions that is dense: for all $f \in L^2$ and $\varepsilon > 0$ there exists $g \in \mathcal{F}$ such that $||f - g||_2 < \varepsilon$. (This also holds in the case when $(X, \mathcal{S}, \mu)$ is infinite and σ-finite.)

4.8. Eigenvalues

Let $(X, \mathcal{S}, \mu)$ be a probability space and let $T : X \to X$ be a measure-preserving transformation. We say that a number $\lambda \in \mathbb{C}$ is an **eigenvalue** of T if there exists a nonzero a.e. function $f \in L^2(X, \mu, \mathbb{C})$

4.8. Eigenvalues

such that
$$f(T(x)) = \lambda f(x) \ \mu\text{-a.e.}$$

The function f is called an **eigenfunction** or **eigenvector** corresponding to λ. Evidently, $\lambda = 1$ is always an eigenvalue. We hope that it will not lead to confusion that we also use λ to represent Lebesgue measure.

Lemma 4.8.1. *If λ is an eigenvalue, then $|\lambda| = 1$, i.e., λ lies in the unit circle in $\mathbb{C}$.*

Proof. Since $f \circ T = \lambda f$ a.e. and $f \in L^2$, using that T is measure-preserving, we have
$$\int |f \circ T|^2 \, d\mu = \int |\lambda|^2 |f|^2 \, d\mu,$$
$$\int |f|^2 \circ T \, d\mu = |\lambda|^2 \int |f|^2 \, d\mu,$$
$$\int |f|^2 \, d\mu = |\lambda|^2 \int |f|^2 \, d\mu.$$

As $\int |f|^2 \, d\mu \neq 0$, $|\lambda|^2 = 1$, so $|\lambda| = 1$. $\square$

Lemma 4.8.2. *Let T be a measure-preserving transformation.*

(1) *If T is ergodic and f is an eigenfunction, then $|f|$ is constant a.e. Therefore, when T is ergodic, we may choose an eigenfunction f with $|f| = 1$ a.e. (i.e., the values of f lie in the unit circle).*

(2) *If λ is an eigenvalue, then the set*
$$E(\lambda) = \{f \in L^2 : f \circ T = \lambda f \, a.e.\}$$
is a vector subspace of L^2.

(3) *T is ergodic if and only if for each eigenvalue λ, the subspace $E(\lambda)$ is 1-dimensional. (In this case λ is called a **simple eigenvalue**.)*

(4) *If f is an eigenfunction corresponding to λ_1, g is an eigenfunction corresponding to λ_2 and $\lambda_1 \neq \lambda_2$, then f and g are orthogonal.*

Proof. For part (1), as f is an eigenfunction, $f \circ T = \lambda f$ a.e., for some $\lambda \in \mathbb{S}^1$. Then $|f| \circ T = |f \circ T| = |\lambda f| = |\lambda||f| = |f|$. The ergodicity of T implies that $|f|$ is constant a.e. Furthermore, we may assume that $|f| \neq 0$ a.e., for if $|f| = 0$ a.e., then $f = 0$ a.e., a contradiction. As $|f|$ is constant a.e., $\frac{f}{|f|}$ is also an eigenfunction, evidently of absolute value 1. Part (2) follows a standard linear algebra proof and is left to the reader. For part (3), suppose that T is ergodic and f is an eigenfunction corresponding to λ. From part (1), $|f| \neq 0$ a.e. Let g be any other eigenfunction corresponding to λ. So the function $h = g/f$ is measurable. One can verify that $h \circ T = h$, therefore $h = c$, a constant a.e. So $g = cf$, showing that $E(\lambda)$ is 1-dimensional. Part (4) also follows the standard linear algebra proof, and is left to the reader. □

By Lemma 4.8.2, when T is ergodic we may assume that eigenfunctions have absolute value 1.

Lemma 4.8.3. *Let T be an ergodic measure-preserving transformation of a Lebesgue probability space $(X, \mathcal{S}, \mu)$. The set $E(T)$ of eigenvalues of T is a countable multiplicative subgroup of the circle group $\mathbb{S}^1$, called the* **eigenvalue group** *of T.*

Proof. If $\lambda_1, \lambda_2 \in E(T)$, then $f_1 \circ T = \lambda f_1, f_2 \circ T = \lambda f_2$ a.e. for some $f_1, f_2 \in L^2$. We may assume that $|f_1| = |f_2| = 1$ a.e. Then $(f_1 \cdot f_2) \circ T = f_1 \circ T \cdot f_2 \circ T = \lambda_1 f_1 \cdot \lambda_2 f_2 = \lambda_1 \lambda_2 \ f_1 \cdot f_2$. Since $|f_1 \cdot f_2| = 1$, $f_1 \cdot f_2 \in L^2$ (in fact, it is in L^∞). Thus, $\lambda_1 \lambda_2 \in E(T)$. Also, $1/f_1(T(x)) = (1/\lambda_1)(1/f_1(x))$ and $1/f_1 \in L^2$. As $1 \in E(T)$ this shows that $E(T)$ is a multiplicative subgroup. To see that it is countable, note that eigenfunctions corresponding to different eigenvalues are orthogonal. If $E(T)$ were uncountable there would exist an uncountable orthonormal subset of $L^2(X, \mathcal{S}, \mu)$, contradicting Lemma 4.7.6. □

An eigenvalue λ is a **rational eigenvalue** if it has finite order, i.e., $\lambda^k = 1$ for some integer $k > 0$. This is the case precisely when $\lambda = \exp(2\pi i r)$ with r a rational number.

4.8. Eigenvalues

Lemma 4.8.4. *Let T be an ergodic measure-preserving transformation. T is totally ergodic if and only if it admits no rational eigenvalues other than 1.*

Proof. Suppose that T is totally ergodic and let λ be an eigenvalue with eigenfunction f. If $\lambda^k = 1$ for some integer k, then $f(T^k(x)) = \lambda^k f(x) = f(x)$. So f must be constant a.e., which implies that $\lambda = 1$.

Suppose now that T^2 is not ergodic. Then there is a nonconstant function g such that $g \circ T^2 = g$ a.e. Define $f = g \circ T - g$. Then $f(T(x)) = g(T^2(x)) - g(Tx) = (-1)(g(T(x)) - g(x)) = (-1)f(x)$. The function f is nonzero a.e. as otherwise g would be constant. Therefore -1 is an eigenvalue of T. If T^k is not ergodic, for an integer $k > 2$, then in a similar way one can show that there is an eigenvalue $\lambda \neq 1$ with $\lambda^k = 1$. The details are left as an exercise. □

A measure-preserving transformation T is said to have **continuous spectrum** if $\lambda = 1$ is its only eigenvalue and it is simple; equivalently, T is ergodic and $\lambda = 1$ is its only eigenvalue. The notation comes from the fact that in this case it can be shown, using the spectral theorem (not covered in this book), that there exists a measure on the circle that is continuous in the sense that it has no atoms other than 1. Evidently, continuous spectrum implies total ergodicity; however, irrational rotations are totally ergodic but do not have continuous spectrum, as the following example shows.

Example 1. Let $R = R_\alpha$ be an irrational rotation on $[0,1)$. Define $f : [0,1) \to \mathbb{S}^1$ by $f(x) = \exp(2\pi i x)$. Then $f(R_\alpha(x)) = f(x+\alpha) = \exp(2\pi i(x+\alpha)) = \exp(2\pi i\alpha)\exp(2\pi i x) = \exp(2\pi i\alpha)f(x)$, showing that $\exp(2\pi i\alpha)$ is an eigenvalue, and so are $\exp(2\pi i k\alpha), k \in \mathbb{Z}$. The reader is asked to show that there are no other eigenvalues.

Example 2. Let T be the dyadic odometer on $[0,1)$. Let $\lambda = \lambda_n = \exp(2\pi i/2^n)$ and define

$$f(x) = \sum_{i=0}^{h_n-1} \lambda^i \, \mathbb{I}_{I_{n,i}}(x).$$

If $x \in I_{n,i}$, then $f(T(x)) = \lambda^{i+1}$ and $f(x) = \lambda^i$, so $f(T(x)) = \lambda f(x)$ for all $x \in X$. As $f \neq 0$, this shows that all 2^n ($n \geq 0$) roots of unity are eigenvalues. Next we show that these are the only eigenvalues. Suppose $g(T(x)) = \beta g(x)$ for some $g \in L^2$ with $|g| = 1$ a.e. For any $\varepsilon > 0$ there is some c so that the set $A = \{x : |g(x) - c| < \varepsilon\}$ has positive measure. Find a level I in some column C_n so that $\lambda(A \cap I) > \frac{3}{4}\lambda(I)$. As $T^{h_n}(I) = I$ and T^{h_n} is measure-preserving, $\lambda(T^{-h_n}(A) \cap A) > 0$, so there is a point $x \in A$ with $T^{h_n}(x) \in A$ and $|g(x)| = 1$. Thus,

$$|g(x) - c| < \varepsilon \quad \text{and} \quad |\beta^{h_n} g(x) - c| = |g(T^{h_n}(x)) - c| < \varepsilon.$$

So, $|g(x) - \beta^{h_n} g(x)| < 2\varepsilon$. Then $|1 - \beta^{h_n}| < 2\varepsilon$. This implies that β is a 2^n-root of 1, since $h_n = 2^n$.

Exercises

(1) Show that if $f : X \to \mathbb{C}$ is a measurable function which is nonzero on a set of positive measure, then for each $\varepsilon > 0$ there exists a constant c such that the set

$$A = \{x : |f(x) - c| < \varepsilon, f(x) \neq 0\}$$

has positive measure.

(2) Complete the proof of Lemma 4.8.2.

(3) Show that if f is an eigenfunction corresponding to λ and $\lambda \neq 1$, then $\int f \, d\mu = 0$ (in this case we say that f belongs to the orthogonal complement of the constants).

(4) Complete the proof of Lemma 4.8.4 as outlined in the proof.

(5) Find the eigenvalue group of a rotation on n points with counting measure.

4.9. Product Measure

This section constructs a measure on the Cartesian product of two measure spaces. This is an extremely useful construction with many applications. We present the detailed construction for the case of the product of two probability spaces, as this is the main application we have in mind. The extension to the product of two σ-finite measure spaces can be obtained with the ideas discussed here and is left as an exercise to the reader.

4.9. Product Measure

The unit square can be seen as the Cartesian product of the unit interval with itself. A question that is raised by this is whether one can construct Lebesgue measure on the square from Lebesgue measure on the interval. In this section we will show how, given two probability spaces $(X, \mathcal{S}(X), \mu)$ and $(Y, \mathcal{S}(Y), \nu)$, one can construct a σ-algebra on $X \times Y$ and a measure on this σ-algebra.

We begin with the construction of a semi-ring on the product space. Let $(X, \mathcal{S}(X), \mu)$ and $(Y, \mathcal{S}(Y), \nu)$ be two probability spaces. Recall that the Cartesian product of X and Y consists of all points of the form (x, y) such that $x \in X$ and $y \in Y$. The first natural subsets to consider are sets of the form $A \times B$ where A is in $\mathcal{S}(X)$ and B is in $\mathcal{S}(Y)$. Sets of this form will be called **measurable rectangles** and the collection of all measurable rectangles is denoted by $\mathcal{S}(X) \times \mathcal{S}(X)$.

Lemma 4.9.1. *Let $(X, \mathcal{S}(X), \mu)$ and $(Y, \mathcal{S}(Y), \nu)$ be two probability spaces. Then the collection $\mathcal{S}(X) \times \mathcal{S}(Y)$ of all measurable rectangles is a semi-ring on $X \times Y$. Furthermore, it contains $X \times Y$.*

Proof. First note that clearly $X \times Y$ is an element of $\mathcal{S}(X) \times \mathcal{S}(Y)$. Now let $A_1 \times B_1$ and $A_2 \times B_2$ be in $\mathcal{S}(X) \times \mathcal{S}(Y)$. Let $A_3 = A_1 \cap A_2$ and $B_3 = B_1 \cap B_2$. Then it easy to verify that $(A_1 \times B_1) \cap (A_2 \times B_2) = A_3 \times B_3$. So $\mathcal{S}(X) \times \mathcal{S}(Y)$ is closed under intersection.

Finally, let $(x, y) \in (A_1 \times B_1) \setminus (A_2 \times B_2)$. If x is in $A_1 \setminus A_2$, then y can be in $B_1 \setminus B_2$ or in $B_1 \cap B_2$. If x is in $A_1 \cap A_2$, then y must be in $B_1 \setminus B_2$. So one can see that

$$\begin{aligned}(A_1 \times B_1) \setminus (A_2 \times B_2) &= (A_1 \setminus A_2) \times (B_1 \setminus B_2) \\ &\sqcup (A_1 \setminus A_2) \times (B_1 \cap B_2) \\ &\sqcup (A_1 \cap A_2) \times (B_1 \setminus B_2),\end{aligned}$$

a finite union of elements of $\mathcal{S}(X) \times \mathcal{S}(X)$. □

To use the theory developed in Section 2.8 we need to construct a countably additive set function on the semi-ring $\mathcal{S}(X) \times \mathcal{S}(Y)$ of measurable rectangles. Define the **product measure** $\mu \times \nu$ on the semi-ring $\mathcal{S}(X) \times \mathcal{S}(Y)$ by

$$\mu \times \nu(A \times B) = \mu(A)\nu(B),$$

for all $A \in \mathcal{S}(X)$ and $B \in \mathcal{S}(Y)$. We should note here that we have not shown yet that $\mu \times \nu$ is a measure. The idea will be to reduce the question to the one-dimensional case. This reduction is obtained by means of the following definition. Given a set E in $X \times Y$ and a point x in E, define the **fiber** at x by

$$E_x = \{y \in Y : (x,y) \in E\}.$$

So, given $A \in \mathcal{S}(X), B \in \mathcal{S}(Y)$,

(4.6) $$(A \times B)_x = \begin{cases} B, & \text{if } x \in A; \\ \emptyset, & \text{if } x \notin A. \end{cases}$$

The following lemma shows that the fibers are measurable.

Lemma 4.9.2. *Let $(X, \mathcal{S}(X), \mu)$ and $(Y, \mathcal{S}(Y), \nu)$ be two probability spaces. Then for every set E in $\mathcal{S}(X) \times \mathcal{S}(Y)$ and every $x \in X$ the fiber set at x, E_x, is in $\mathcal{S}(Y)$.*

Now we are ready for the main technical lemma of this section.

Lemma 4.9.3. *Let $(X, \mathcal{S}(X), \mu)$ and $(Y, \mathcal{S}(Y), \nu)$ be two probability spaces. Then $\mu \times \nu$ is a countably additive probability set function on $\mathcal{S}(X) \times \mathcal{S}(Y)$.*

Proof. The main property that needs to be verified is that $\mu \times \nu$ is countably additive (clearly, $\mu \times \nu$ is nonnegative, assigns 0 to the empty set, and $\mu \times \nu(X \times Y) = 1$). So let $\{A_i \times B_i\}$ be a countable disjoint collection of sets in $\mathcal{S}(X) \times \mathcal{S}(Y)$ whose union is in $\mathcal{S}(X) \times \mathcal{S}(Y)$. Write

$$A \times B = \bigsqcup_{i=1}^{\infty} A_i \times B_i,$$

for $A \in \mathcal{S}(X), B \in \mathcal{S}(Y)$. We know that $(A \times B)_x, (A_i \times B_i)_x$ are in $\mathcal{S}(Y)$ for each $x \in X$. By (4.6),

$$\nu[(A \times B)_x] = \mathcal{I}_A(x)\nu(B), \quad \nu[(A_i \times B_i)_x] = \mathcal{I}_{A_i}(x)\nu(B_i), \quad i \geq 1.$$

Thus,

$$\mathcal{I}_A(x)\nu(B) = \sum_{i=1}^{\infty} \mathcal{I}_{A_i}(x)\nu(B_i).$$

4.9. Product Measure

Integrating both sides with respect to μ and using Corollary 4.4.3,

$$\mu(A)\nu(B) = \int \sum_{i=1}^{\infty} \mathcal{I}_{A_i}(x)\nu(B) \, d\mu$$

$$= \sum_{i=1}^{\infty} \int \mathcal{I}_{A_i}(x)\nu(B_i) \, d\mu$$

$$= \sum_{i=1}^{\infty} \mu(A_i)\nu(B_i).$$

Therefore, $\mu \times \nu$ is countably additive. $\square$

Before stating the theorem, we introduce a measurable structure on the product space, and for this the natural choice is to consider the σ-algebra generated by the measurable rectangles. Given probability spaces $(X, \mathcal{S}(X), \mu)$ and $(Y, \mathcal{S}(Y), \nu)$, define $\mathcal{S}(X) \star \mathcal{S}(Y)$ to be the σ-algebra generated by the measurable rectangles. We should mention that this notation is not standard and some authors use $\mathcal{S}(X) \times \mathcal{S}(Y)$ or $\mathcal{S}(X) \otimes \mathcal{S}(Y)$ for the σ-algebra generated by the measurable rectangles.

Theorem 4.9.4. *Let $(X, \mathcal{S}(X), \mu)$ and $(Y, \mathcal{S}(Y), \nu)$ be two probability spaces. Define $\mu \times \nu$ on $\mathcal{S}(X) \times \mathcal{S}(Y)$ by*

$$\mu \times \nu(A \times B) = \mu(A)\nu(B)$$

for $A \in \mathcal{S}(X)$ and $B \in \mathcal{S}(Y)$. Then $\mu \times \nu$ has a unique extension to a probability measure, also denoted $\mu \times \nu$, defined on $\mathcal{S}(X) \star \mathcal{S}(Y)$, the σ-algebra generated by $\mathcal{S}(X) \times \mathcal{S}(Y)$.

Proof. As we already know that $\mu \times \nu$ is a countably additive finite set function on $\mathcal{S}(X) \times \mathcal{S}(Y)$, Theorem 2.8.2 shows that there is a unique extension to a probability measure on the generated σ-algebra. $\square$

We note that as explained in Section 2.8 for the case of this type of constructions, the semi-ring of measurable rectangles is a sufficient semi-ring for the extension measure, and therefore the approximation results of Section 2.7 hold.

As an application we introduce the notion of the Cartesian product transformation. Let T be a transformation defined on the probability space $(X, \mathcal{S}(X), \mu)$ and let S be a transformation defined on

the probability space $(Y, \mathcal{S}(Y), \nu)$. Then we define a transformation $T \times S$ on the product space $(X \times Y, \mathcal{S}(X) \star \mathcal{S}(Y), \mu \times \nu)$ by
$$T \times S(x, y) = (T(x), S(y)).$$
It can be shown that if $\mathcal{C}$ is a sufficient semi-ring (or algebra) for X, then $\{I \times J : I, J \in C\}$ is a sufficient semi-ring (or algebra) for $X \times Y$. From this it follows that if T and S are measure-preserving, then so is $T \times S$.

Exercises

(1) Let $(X, \mathcal{S}(X), \mu)$ and $(Y, \mathcal{S}(Y), \nu)$ be two probability spaces. Show that the collection of sets $\mathcal{A} = \{E \in S(X) \otimes S(Y) : E_x \in \mathcal{S}(Y) \text{ for all } x \in X\}$ is a σ-algebra.

(2) Let (X, d) and (Y, q) be two separable, complete metric spaces. Show that $\mathcal{B}(X \times Y) = \mathcal{B}(X) \star \mathcal{B}(Y)$.

(3) Let T be the dyadic odometer. Show that $T \times T$ is not ergodic for the product measure.

(4) Show that $R_\alpha \times R_\alpha$ is not ergodic for 2-dimensional Lebesgue measure. When is $R_\alpha \times R_\beta$ ergodic?

* (5) Let T be the dyadic odometer and let R be an irrational rotation. Is the product transformation $T \times R$ ergodic?

* (6) Let T and S be finite measure-preserving transformations. Show that if $T \times S$ is ergodic, then $E(T) \cap E(S) = \{1\}$. (The converse is also true but harder to show.)

Chapter 5

The Ergodic Theorem

This chapter is devoted to the ergodic theorem and several of its applications. The ergodic theorem, proved in 1931, first by J. von Neumann in the case of convergence in the mean and then by G. D. Birkhoff in the stronger case of pointwise convergence, solved an important question that arose in statistical mechanics: to give a condition under which the time-average and the space-average of a dynamical system agree. The pointwise convergence version of the ergodic theorem, known as Birkhoff's ergodic theorem, in its simplest form asserts that a finite measure-preserving transformation T on a probability space $(X, \mathcal{S}, \mu)$ is ergodic if and only if

$$(5.1) \qquad \lim_{N \to \infty} \frac{1}{N} \sum_{n=0}^{N-1} \mathbb{I}_A(T^n(x)) = \mu(A)$$

holds for each measurable set A in X and for each point x of X outside of a set $N = N(A)$ of measure zero (in general depending on A).

The quantity on the left of (5.1) is the limit of the average number of times that images of x under iterates of T land in A. We think of this quantity as the time average of visits to A by the orbit of x. The quantity on the right of (5.1) can be thought of as the space average of A. The Birkhoff ergodic theorem implies that for any measurable set A and for any point x outside some set N of measure zero, the time-average is asymptotically equal to the space-average. The theorem holds in a more general case. First, the function $\mathbb{I}_A$ may be replaced by any integrable function f, and in this case $\mu(A)$ is replaced by the integral of f. Also, there is a version of the theorem that does not require T to be ergodic, just measure-preserving, but the conclusion is weaker in this case. This theorem has had many generalizations over the years, some of which we discuss in later sections and others in the Bibliographical Notes.

5.1. The Birkhoff Ergodic Theorem

In this section, T will be a fixed measure-preserving transformation of a finite measure space $(X, \mathcal{S}, \mu)$. The most important case in this section is when $\mu(X) = 1$.

To illustrate the kind of limits we are concerned with, we start by recalling Application 1 from Section 3.2. The question that remained to be answered was: what is the frequency of appearances of 7 as the first digit in the powers $2^n, n \geq 0$? We saw that 2^n starts with a 7 if and only if $R_{\log 2}^n(0) \in [\log 7, \log 8)$ (where log is base 10). Then the frequency

$$\lim_{n \to \infty} \frac{1}{n} \#\{i : 0 \leq i < n, 2^i \text{ starts with a } 7\}$$

can be written as

$$(5.2) \qquad \lim_{n \to \infty} \frac{1}{n} \sum_{i=0}^{n-1} \mathbb{I}_A(R_{\log 2}^i(0)),$$

where $A = [\log 7, \log 8)$. *A priori* there is no reason to think that the limit in (5.2) exists. Thus, the question that is left to answer is whether the limit in (5.2) exists, and if so, what its value is. The ergodic theorem asserts that when T is ergodic and finite measure-preserving on a probability space, limits such as (5.2) exist and equal the measure of A for all points x in X outside a set of measure zero (in general, depending on A). The reader will note that the limit in (5.2) is taken at the particular point $x = 0$. In general, knowing that the limit exists a.e. does not imply that it will exist for a particular point such as $x = 0$; however, because of the nature of irrational rotations one can show that in this case the limit exists everywhere, and in particular at $x = 0$. Thus one can use the ergodic theorem and the

5.1. The Birkhoff Ergodic Theorem

nature of irrational rotations to show that the limit (5.2) exists and equals the measure of A. The expression in (5.2) represents the time average for the visits of the orbit of 0, under the transformation $R_{\log 2}$, to the set A. As mentioned before, the study of these averages came from considerations in statistical mechanics and are what started our subject.

Rather than studying the averages only for characteristic functions, we will consider them for arbitrary integrable functions. We start with the statement of the theorem.

Theorem 5.1.1 (Birkhoff Ergodic Theorem). *Let $(X, \mathcal{S}, \mu)$ be a measure space and let T be a measure-preserving transformation on $(X, \mathcal{S}, \mu)$. If $f : X \to \mathbb{R}$ is an integrable function, then*

(1) $\lim_{n \to \infty} \frac{1}{n} \sum_{i=0}^{n-1} f(T^i(x))$ *exists for all $x \in X \setminus N$, for some null set N depending on f. Denote this limit by $\tilde{f}(x)$.*

(2) $\tilde{f}(Tx) = \tilde{f}(x)$ *a.e.*

(3) *For any measurable set A that is T-invariant,*

$$\int_A f \, d\mu = \int_A \tilde{f} \, d\mu.$$

In particular, if T is ergodic, then

(5.3) $$\lim_{n \to \infty} \frac{1}{n} \sum_{i=0}^{n-1} f(T^i(x)) = \int f \, d\mu \quad a.e.$$

We shall present two proofs of this theorem. This first one is in the case when T is ergodic and the proof will yield at the same time the existence of the limit and its identification as the integral of f. The second proof will not assume that T is ergodic and must show separately that the limit exists a.e. and that is satisfies (3) above. In both cases the hard part of the proof is to show the existence of the limit. It is straightforward to show that if the limit exists, then it must be a T-invariant function; this is done in the lemma below. Also, when T is ergodic, using the Lebesgue convergence theorems it is not hard to show that, if the limits exists, then it must be the integral of f (see Exercise 4).

We introduce some useful notation to be used throughout this Section. For any function f define

$$(5.4) \qquad f_n(x) = \sum_{i=0}^{n-1} f(T^i(x)), \quad n \geq 1,$$

$$(5.5) \qquad f_*(x) = \liminf_{n \to \infty} \frac{1}{n} \sum_{i=0}^{n-1} f(T^i(x)), \text{ and}$$

$$(5.6) \qquad f^*(x) = \limsup_{n \to \infty} \frac{1}{n} \sum_{i=0}^{n-1} f(T^i(x)).$$

Lemma 5.1.2. *The functions $f_*(x)$ and $f^*(x)$ are T-invariant.*

Proof. The reader is asked to verify that the following equality follows from the definitions:

$$(5.7) \qquad \frac{1}{n} f_n(T(x)) = \frac{n+1}{n} \frac{1}{n+1} f_{n+1}(x) - \frac{1}{n} f(x).$$

Next observe that $\lim_{n \to \infty} \frac{1}{n} f(x) = 0$ and $\lim_{n \to \infty} \frac{n+1}{n} = 1$. Taking lim inf and lim sup of both sides of (5.7) shows that $f_*(T(x)) = f_*(x)$ and $f^*(T(x)) = f^*(x)$. □

The invariance of the limit $\lim_{n \to \infty} f_n(x)$, when it exists, is now a direct consequence of Lemma 5.1.2. Our task is now to show the existence of this limit. To this end, we start with a technical lemma. Its use will become apparent later when we choose a particular function τ. As we shall see, to show the existence of the limit we need to establish some inequalities, and the following lemma will give us the basic inequality in (5.9) when it is applied for a particular choice of f and τ.

Lemma 5.1.3. *Let g be a measurable function and let $p \geq 1$ be an integer. If there exists a function $\tau : X \to \{1, 2, \ldots, p\}$ such that*

$$g_{\tau(x)}(x) \leq 0 \quad \text{for all } x \in X,$$

then for all $n \geq p$,

$$(5.8) \qquad g_n(x) \leq \sum_{i=n-p}^{n-1} |g(T^i(x))| \quad \text{for all } x \in X.$$

5.1. The Birkhoff Ergodic Theorem

Proof. The hypothesis states that if we sum the result of evaluating g at consecutive terms of the orbit of x, from x to $T^{\tau(x)-1}(x)$, we obtain a number that is ≤ 0. So, if $\tau(x) < n$, we can then write that

$$g_n(x) \leq \sum_{i=\tau(x)}^{n-1} g(T^i(x)).$$

Define a *block* to be a finite set of ordered values consisting of g evaluated at a consecutive sequence of elements of the positive orbit of x. Let us call a block *good* if the sum of all of the terms in that block is ≤ 0. We just saw that the first good block starts at $g(x)$ and ends at $g(T^{\tau(x)-1}(x))$, as $g_{\tau(x)}(x) \leq 0$ by assumption. We quickly observe that there may be other good blocks, and so more terms that can be discarded. The next good block starts at $y = T^{\tau(x)}(x)$ and ends at $T^{\tau(y)-1}(y)$, provided that $\tau(y) < n$. So we continue discarding good blocks until it is not possible anymore, and it is not possible when there are not enough remaining terms to sum to a good block. But this number of remaining terms must be at most p, as that is the bound of τ. It follows that $g_n(x)$ must be bounded above by the absolute value of the last p remaining terms, so we obtain (5.8). □

The following lemma is used in a fundamental way in the proof of the Birkhoff ergodic theorem.

Lemma 5.1.4. *Let f be an integrable function and let $p \geq 1$ be an integer.*

(1) *Define*

$$E_p = \{x \in X : f_n(x) \geq 0 \text{ for all } n, 1 \leq n \leq p\}.$$

Then for all integers $n \geq p$ and for a.e. $x \in X$,

(5.9) $$f_n(x) \leq \sum_{i=0}^{n-1} f(T^i(x))\mathbb{I}_{E_p}(T^i(x)) + \sum_{i=n-p}^{n-1} |f(T^i(x))|.$$

(2) *For each real number r define*

$$E_p^r = \{x \in X : \frac{1}{n}\sum_{i=0}^{n-1} f(T^i(x)) \geq r \text{ for all } 1 \leq n \leq p\}.$$

Then

(5.10) $$\int f\,d\mu \leq \int_{E_p^r} f\,d\mu + r(1-\mu(E_p^r)).$$

Proof. We will apply Lemma 5.1.3 to the function

$$g(x) = f(x) - f(x)\mathbb{I}_{E_p}(x).$$

We first need a function τ that satisfies the hypothesis of Lemma 5.1.3 for g. We claim that the following function τ will work for g. Define

$$\tau(x) = \begin{cases} 1 & \text{if } x \in E_p; \\ \min\{1 \leq k \leq p : f_k(x) < 0\} & \text{if } x \notin E_p. \end{cases}$$

We first show that

(5.11) $$f_{\tau(x)}(x) \leq \sum_{i=0}^{\tau(x)-1} f(T^i(x))\mathbb{I}_{E_p}(T^i(x)) \text{ a.e.}$$

To show (5.11), we first note that if $x \in E_p$, then $\tau(x) = 1$ and also $f(x)\mathbb{I}_{E_p}(x) = f(x)$, so we obtain an equality. Now when $x \notin E_p$ then $f_{\tau(x)}(x) < 0$. In this case, it suffices to show that each term in the sum on the right-hand side of (5.11) is nonnegative. Note that the ith term will be 0 when $T^i(x) \notin E_p$. When $T^i(x) \in E_p$ then $f(T^i(x)) = f_1(T^i(x)) \geq 0$. This completes the proof of inequality (5.11).

Next, we note that it follows from the definition of g that

$$g_{\tau(x)}(x) = f_{\tau(x)}(x) - \sum_{i=0}^{\tau(x)-1} f(T^i(x))\mathbb{I}_{E_p}(T^i(x)),$$

and thus by (5.11) $g_\tau \leq 0$. Therefore we can apply Lemma 5.1.3 to obtain

$$g_n(x) \leq \sum_{i=n-p}^{n-1} |g(T^i(x))|,$$

5.1. The Birkhoff Ergodic Theorem

which means that

$$f_n(x) - \sum_{i=0}^{n-1} f(T^i(x))\mathbb{I}_{E_p}(T^i(x))$$
$$\leq \sum_{i=n-p}^{n-1} |f(T^i(x)) - f(T^i(x))\mathbb{I}_{E_p}(T^i(x))|.$$

Finally, observe that

$$\sum_{i=n-p}^{n-1} |f(T^i(x)) - f(T^i(x))\mathbb{I}_{E_p}(T^i(x))| \leq \sum_{i=n-p}^{n-1} |f(T^i(x))|,$$

and therefore

$$f_n(x) \leq \sum_{i=0}^{n-1} f(T^i(x))\mathbb{I}_{E_p}(T^i(x)) + \sum_{i=n-p}^{n-1} |f(T^i(x))|,$$

which completes the proof of the first part of the lemma.

For the proof of part (2), note that

$$E_p^r = \{x : (f - r)_n(x) \geq 0, \text{ for all } 1 \leq n \leq p\}.$$

Then by the first part applied to the function $f - r$,

$$(f - r)_n(x) \leq \sum_{i=0}^{n-1} (f - r)(T^i(x)) \, \mathbb{I}_{E_p^r}(T^i(x)) + \sum_{i=n-p}^{n-1} |(f - r)(T^i(x))|.$$

Integrating both sides of the inequality and using that T is measure-preserving obtains

(5.12) $$n \int f - r \, d\mu \leq n \int_{E_p^r} f - r \, d\mu + \sum_{i=n-p}^{n-1} \int |f - r| \, d\mu.$$

As the sum on the right-hand side has at most p terms, after dividing by n and taking limits as $n \to \infty$, we have

$$\int f - r \, d\mu \leq \int_{E_p^r} f - r \, d\mu,$$

which gives (5.10). □

We are now ready to prove our main theorem.

Proof of Theorem 5.1.1. We will prove the theorem under the assumption that T is ergodic. First we show that for all integrable functions f,

(5.13) $$\int f \, d\mu \leq f_*(x) \text{ a.e.}$$

Write
$$A = \{x : f_*(x) < \int f \, d\mu\}.$$

To show (5.13) it suffices to show that $\mu(A) = 0$. So suppose that $\mu(A) > 0$. One can write
$$A = \bigcup_{r \in \mathbb{Q}} \{x : f_*(x) < r < \int f \, d\mu\}.$$

Let

(5.14) $$C_r = \{x : f_*(x) < r < \int f \, d\mu\}.$$

Then there is some rational number r such that $\mu(C_r) > 0$. We claim that C_r is T-invariant modμ. This follows from Lemma 5.1.2 and the calculation is left to the reader as an exercise. As T is ergodic, then $\mu(C_r) = 1$. Since

$$E_p^r = \{x : \frac{1}{n} \sum_{i=0}^{n-1} f(T^i(x)) \geq r, \text{ for all } 1 \leq n \leq p\},$$

then the fact that $\mu(C_r) = 1$ implies

$$\mu(\bigcap_{p=1}^{\infty} E_p^r) = 0.$$

So
$$\lim_{p \to \infty} \mu(E_p^r) = 0,$$

and therefore by (5.10)
$$\int f \, d\mu \leq r,$$

5.1. The Birkhoff Ergodic Theorem

which contradicts our choice of r; therefore $\mu(A) = 0$. This completes the proof of (5.13). Now apply (5.13) to $-f$ to obtain

$$\int -f\,d\mu \leq \liminf_{n\to\infty} \frac{1}{n} \sum_{i=0}^{n-1} -f(T^i(x)) \text{ a.e., or}$$

$$\int f\,d\mu \geq \limsup_{n\to\infty} \frac{1}{n} \sum_{i=0}^{n-1} f(T^i(x)) \text{ a.e.}$$

This together with (5.13) implies

$$\int f\,d\mu = \lim_{n\to\infty} \frac{1}{n} \sum_{i=0}^{n-1} f(T^i(x)) \text{ a.e.,}$$

which completes the proof of the theorem. □

As an application we prove a useful and interesting characterization of ergodicity. We should note, however, that the theorem below can also be obtained for the von Neumann ergodic theorem.

Theorem 5.1.5. *Let T be a finite measure-preserving transformation on a probability space $(X, \mathcal{S}, \mu)$. T is ergodic if and only if for all measurable sets A, B,*

(5.15) $$\lim_{n\to\infty} \frac{1}{n} \sum_{i=0}^{n-1} \mu(T^{-i}(A) \cap B) = \mu(A)\mu(B).$$

Proof. Suppose that T is ergodic and let A, B be measurable sets. Then $\mathbb{I}_A$ is integrable and, by the Birkhoff ergodic theorem,

$$\lim_{n\to\infty} \frac{1}{n} \sum_{i=0}^{n-1} \mathbb{I}_A(T^i(x)) = \mu(A) \text{ a.e.}$$

Then

$$\lim_{n\to\infty} \frac{1}{n} \sum_{i=0}^{n-1} \mathbb{I}_A(T^i(x))\mathbb{I}_B(x) = \mu(A)\mathbb{I}_B(x) \text{ a.e.}$$

We note that for all $n > 0$,

$$\left| \frac{1}{n} \sum_{i=0}^{n-1} \mathbb{I}_A(T^i(x))\mathbb{I}_B(x) \right| \leq 1 \text{ a.e.}$$

So by the Dominated Convergence Theorem (Theorem 4.6.3),

$$\lim_{n\to\infty} \int \frac{1}{n} \sum_{i=0}^{n-1} \mathbb{I}_A(T^i(x))\mathbb{I}_B(x)d\mu(x)$$

$$= \int \lim_{n\to\infty} \frac{1}{n} \sum_{i=0}^{n-1} \mathbb{I}_A(T^i(x))\mathbb{I}_B(x)d\mu(x)$$

$$= \int \mu(A)\mathbb{I}_B(x)d\mu(x)$$

$$= \mu(A)\mu(B).$$

Next, we note that

$$\int \frac{1}{n} \sum_{i=0}^{n-1} \mathbb{I}_A(T^i(x))\mathbb{I}_B(x)d\mu(x) = \frac{1}{n} \sum_{i=0}^{n-1} \int \mathbb{I}_{T^{-i}(A)\cap B}(x)d\mu(x)$$

$$= \frac{1}{n} \sum_{i=0}^{n-1} \mu(T^{-i}(A) \cap B).$$

Therefore,

$$\lim_{n\to\infty} \frac{1}{n} \sum_{i=0}^{n-1} \mu(T^{-i}(A) \cap B) = \mu(A)\mu(B).$$

For the converse, note that if A is a T-invariant set, then

$$\frac{1}{n} \sum_{i=0}^{n-1} \mu(T^{-i}(A) \cap A) = \mu(A),$$

but by (5.15) with $B = A$,

$$\lim_{n\to\infty} \frac{1}{n} \sum_{i=0}^{n-1} \mu(T^{-i}(A) \cap A) = \mu(A)\mu(A).$$

So $\mu(A) = \mu(A)^2$, which means $\mu(A)\mu(A^c) = 0$ and T is ergodic. □

The remainder of this section presents another proof of the ergodic theorem and may be omitted on a first reading. It is based on the maximal ergodic theorem (Lemma 5.1.7) and in this case our proof does not assume that T is ergodic; we also do not include as many details as in the previous argument. We start with a lemma similar to Lemma 5.1.4.

5.1. The Birkhoff Ergodic Theorem

Lemma 5.1.6. *Let f be a measurable function and let $p \geq 1$ be an integer. Define*

$$G_p = \{x \in X : f_n(x) > 0 \text{ for some } n, 1 \leq n \leq p\}.$$

Then for all integers $n \geq p$ and for $x \in X$,

$$(5.16) \qquad 0 \leq \sum_{i=0}^{n-1} f(T^i(x))\mathbb{I}_{G_p}(T^i(x)) + \sum_{i=n-p}^{n-1} |f(T^i(x))|.$$

Proof. We apply Lemma 5.1.3 to the function

$$g(x) = -f(x)\mathbb{I}_{G_p}(x).$$

First define a function τ by

$$\tau(x) = \begin{cases} 1 & \text{if } x \notin G_p; \\ \min\{1 \leq k \leq p : f_k(x) > 0\} & \text{if } x \in G_p. \end{cases}$$

To show that τ satisfies the hypothesis of Lemma 5.1.3 for g it suffices to verify that

$$(5.17) \qquad 0 \leq \sum_{i=0}^{\tau(x)-1} f(T^i(x))\mathbb{I}_{G_p}(T^i(x)) \text{ a.e.}$$

For this observe that if $x \in G_p$, then $0 < f_\tau = \sum_{i=0}^{\tau-1} f \circ T^i$. To complete this part of the proof the reader is asked to show that $T^i(x) \in G_p$ for $0 \leq i \leq \tau(x)$. Finally, apply Lemma 5.1.3. $\square$

Lemma 5.1.7 (Maximal Ergodic Theorem). *Let f be an integrable function and define*

$$G(f) = \{x \in X : f_n(x) > 0 \text{ for some } n > 0\}.$$

Then

$$(5.18) \qquad \int_{G(f)} f \geq 0.$$

Proof. Let $p \geq 1$ be an integer and define

$$G_p = \{x \in X : f_n(x) > 0 \text{ for some } n, 1 \leq n \leq p\}.$$

Integrating both sides of inequality (5.16) and using that T is measure-preserving gives

$$(5.19) \qquad 0 \leq n \int_{G_p} f \, d\mu + \sum_{i=n-p}^{n-1} \int |f| \, d\mu.$$

After dividing by n and taking limits as $n \to \infty$ we obtain

$$(5.20) \qquad \int_{G_p} f \geq 0$$

for each $p \geq 1$. Next we observe that $G(f) = \bigcup_{p>0} G_p$. An application of the dominated convergence theorem completes the proof. $\square$

Second proof of Theorem 5.1.1. First we show the existence of the limit. For numbers r and s write

$$A(r,s) = \{x : f_*(x) < r < s < f^*(x)\}.$$

If we show that $\mu(A(r,s)) = 0$ for all r, s, then it follows that $f_* = f^*$ a.e. By Lemma 5.1.2 the sets $A(r,s)$ are T-invariant. So we can apply Lemma 5.1.7 to T restricted to $A(r,s)$ and the set $G(f - s)$. As $A(r,s) \cap G(f - s) = A(r,s)$ we obtain

$$\int_{A(r,s)} f \, d\mu \geq s\mu(A(r,s)).$$

Applying the same result to $G(r - f)$ gives $r\mu(A(r,s)) \geq \int_{A(r,s)} f \, d\mu$. If $\mu(A(r,s)) > 0$, then these two inequalities yield that $s \leq r$, a contradiction. Therefore $f_* = f^*$ a.e.

To show part (3), writing $f = f^+ - f^-$, it suffices to consider the case when f is nonnegative. First assume that f is a bounded nonnegative function a.e. Then the dominated convergence theorem implies

$$\int_A \tilde{f} \, d\mu = \int_A \lim_{n \to \infty} \frac{1}{n} \sum_{i=0}^{n-1} f(T^i(x)) \, d\mu$$

$$= \lim_{n \to \infty} \frac{1}{n} \sum_{i=0}^{n-1} \int_A f(T^i(x)) \, d\mu = \int_A f \, d\mu,$$

5.1. The Birkhoff Ergodic Theorem

for any T-invariant set A. The case for nonnegative f in L^1 follows by an approximation argument. First we note that Fatou's lemma implies that

$$||\tilde{f}||_1 = \int |\lim_{n\to\infty} \frac{1}{n} f_n(x)| \, d\mu$$
$$\leq |\liminf_{n\to\infty} \int \frac{1}{n} f_n(x)| \, d\mu = ||f||_1.$$

Now approximate f with a bounded function g and consider

$$|\int_A f \, d\mu - \int_A \tilde{f} \, d\mu| \leq |\int_A f \, d\mu - \int_A g \, d\mu| + |\int_A g \, d\mu - \int_A \tilde{f} \, d\mu|$$
$$\leq \int_A |f - g| \, d\mu + |\int_A g \, d\mu - \int_A \tilde{g} \, d\mu|$$
$$+ |\int_A \tilde{g} \, d\mu - \int_A \tilde{f} \, d\mu|$$
$$\leq ||f - g||_1 + 0 + ||f - g||_1.$$

As $||f - g||_1$ can be made arbitrarily small (see Exercise 4.7.10), this completes the proof of the theorem.

(We note that a similar argument using the dominated convergence theorem shows that $\frac{1}{n} f_n$ converges to f in L^1. This will be treated in more detail in Section 5.4.) □

Exercises

(1) Let $T: X \to X$ be a transformation. Establish the equality

$$\frac{1}{n} \sum_{i=0}^{n-1} f(T^i(Tx)) = [\frac{n+1}{n} \frac{1}{n+1} \sum_{i=0}^{n} f(T^i(x)) - \frac{f(x)}{n}].$$

(2) Let $T: X \to X$ be a measure-preserving transformation with $\mu(X) = 1$. Show that if for every measurable set A the limit

$$\lim_{n\to\infty} \frac{1}{n} \sum_{i=0}^{n-1} \mathbb{I}_A(T^i(x))$$

exists and equals $\mu(A)$ a.e., then T must be ergodic.

(3) Let $T : X \to X$ be a measure-preserving transformation with $\mu(X) = 1$. Without assuming the ergodic theorem, show that if for every measurable set A the limit
$$\lim_{n \to \infty} \frac{1}{n} \sum_{i=0}^{n-1} \mathbb{I}_A(T^i(x))$$
exists and is constant a.e., then it must equal $\mu(A)$.

(4) Extend Exercise 3 when $\mathbb{I}_A$ is replaced by a bounded integrable function f.

(5) Show that C_r in (5.14) is T-invariant mod μ.

(6) For a more formal proof of Lemma 5.1.3, define inductively an increasing sequence of "times" by
$$\tau_0(x) = 0;$$
$$\tau_{m+1}(x) = \tau(T^{\tau_m(x)}(x)) + \tau_m(x).$$
Show that
$$g_n(x) \le g_{n-\tau_m(x)}(T^{\tau_m(x)}(x)) = \sum_{i=\tau_m(x)}^{n-1} g(T^i(x))$$
$$\le \sum_{i=n-p}^{n-1} |g(T^i(x))|,$$
where $m \ge 1$ is the first integer satisfying $\tau_m(x) < n \le \tau_{m+1}(x)$.

∗ (7) Modify the proof of the ergodic theorem to show that if T is an infinite measure-preserving ergodic transformation, then $\lim_{n \to \infty} \frac{1}{n} \sum_{i=0}^{n-1} f(T^i(x)) = 0$ a.e. for all integrable functions f. What can you conclude if T is not ergodic?

5.2. Normal Numbers

Recall that every number x in $[0,1]$ has a representation in binary form $x = \sum_{i=1}^{\infty} \frac{a_i}{2^i}$, where $a_i \in \{0,1\}$. We define in this section an important notion introduced by Borel in 1909. A number $x \in [0,1]$ is said to be **simply normal to base** 2 if in its binary expansion the frequency of appearances of 0 is $\frac{1}{2}$ (and so is the frequency of

5.2. Normal Numbers

appearances of 1). For example, 1/3 is simply normal to base 2. (For numbers that have two binary representations the reader may verify that the notion of simply normal is independent of the representation.) However, 111, say, does not appear in the binary expansion of 1/3 with any frequency. A number x is said to be **normal to base 2** if in its binary expansion the frequency of appearances of any finite sequence of digits $a_1 \ldots a_k$ (where a_i, $1 \leq i \leq k$, is 0 or 1) is 2^{-k}. Normal numbers to other bases are defined in a similar manner.

We now use a transformation to study the frequency of binary representations of digits. Let $T(x) = 2x \pmod{1}$. We saw in Theorem 3.7.6 that T is a finite measure-preserving ergodic transformation of $([0, 1), \mathfrak{L}, \lambda)$. Note that x is simply normal to base 2 if

$$\lim_{n \to \infty} \frac{1}{n} \sum_{i=0}^{n-1} \mathbb{I}_{[0,1/2)}(T^i(x)) = \frac{1}{2}.$$

Now we observe how T behaves on x when we use its binary representation. If $x = \sum_{i=1}^{\infty} \frac{a_i}{2^i}$, then

$$T(x) = \sum_{i=1}^{\infty} \frac{a_{i+1}}{2^i}.$$

We can think of T as a shift transformation since it takes the representation $(a_1, a_2, a_3, \ldots)$ to the representation $(a_2, a_3, a_4, \ldots)$. With this interpretation we can use T to study the digits in the representation of x. For example, $T^{i-1}(x) \in [0, 1/2)$, $i \geq 1$, if and only if the ith digit in the binary representation of x is 0 (similarly, $T^{i-1}(x) \in [1/2, 1)$ if and only if the ith digit in the binary representation of x is 1). Furthermore, if $T^{i-1}(x) \in [1/4, 1/2)$ then the two digits starting in position i in the binary representation of x are 0 followed by 1. Now let $D_{k,\ell} = [\frac{\ell}{2^k}, \frac{\ell+1}{2^k})$ denote the ℓth dyadic interval of order k ($k \geq 1$, $\ell = 1, \ldots, 2^k - 1$).

The reader is asked to verify that $T^i(x) \in D_{k,\ell}$ if and only if the k digits in the binary representation of x starting at position $i-1$ are the same as the k digits in the binary representation of $\frac{\ell}{2^k}$ that ends with 0's. (For example, the first two digits in the binary representation of $\frac{1}{4}$ that ends with 0's are 0 and 1.) So x is normal to base 2 if and only if for all $k \geq 1$, $\ell = 0, \ldots, 2^k - 1$,

$$\lim_{n \to \infty} \frac{1}{n} \sum_{i=0}^{n-1} \mathbb{I}_{D_{k,\ell}}(T^i(x)) = \frac{1}{2^k}.$$

We are now ready to prove the following interesting theorem. It is a special case of what is known as the Law of Large Numbers. It can be interpreted as saying that if we pick a number at random (according to Lebesgue measure) the frequency of appearances of a word of length k of 0's and 1's in its binary representation is 2^{-k}.

Theorem 5.2.1 (Borel's Normal Number Theorem). *Almost every number in $[0, 1]$ is a normal number to base 2.*

Proof. Let x be in $[0, 1]$. By the ergodic theorem,

(5.21) $$\lim_{n \to \infty} \frac{1}{n} \sum_{i=0}^{n-1} \mathbb{I}_{D_{k,\ell}}(T^i(x)) = \lambda(D_{k,\ell}) = 2^{-k}$$

for all x outside a null set $N_{k,\ell}$. As the collection of null sets $N_{k,\ell}$ is countable, there is a null set N such that for all $x \in [0, 1] \setminus N$ the limit in (5.21) holds. □

One can define the notion of normality for any number x in $\mathbb{R}$ by just considering x (mod 1), and this notion can be defined with respect to any base. An important open question is whether π is normal or even simply normal. In 1933 Champernowne proved that the number $0.123456789101112\ldots$ is normal to base 10.

Exercises

(1) Show that almost every number x in $\mathbb{R}$ is normal to base 2.

(2) Show that almost every number x in $[0, 1]$ is normal in base 10.

(3) Construct an infinite number of simply normal numbers to base 10.

(4) A number is **absolutely normal** if it is normal to any (integer) base b. Show that almost all numbers are absolutely normal.

5.3. Weyl Equidistribution

(5) Show that a rational number cannot be absolutely normal.

(6) Show that the set of normal numbers is a Borel set.

5.3. Weyl Equidistribution

A sequence $\{a_i\}$ in $[0,1]$ is said to be **equidistributed** or **uniformly distributed** in $[0,1]$ if for all intervals I in $[0,1]$ it is the case that

$$\lim_{n\to\infty} \frac{1}{n}\#\{i: 0 \le i < n, a_i \in I\} = \lambda(I).$$

In this section, we show that for any irrational number α the sequence $n\alpha \pmod{1}$ is uniformly distributed in $[0,1]$. Note that $n\alpha \pmod{1} = R^n(0)$. This will also answer the remaining questions on Gelfand's question on the frequency of appearances of 7 in powers of 2 in Section 3.2.

Theorem 5.3.1. *Let α be an irrational number and $R = R_\alpha$. Then for every interval I in $[0,1)$,*

$$\lim_{n\to\infty} \frac{1}{n} \sum_{i=0}^{n-1} \mathbb{I}_I(R^i(x)) = \lambda(I) \quad \text{for all } x \in [0,1).$$

Proof. As R is ergodic, by the ergodic theorem, given an interval I there is a nullset $N(I)$ so that for all $x \in [0,1) \setminus N(I)$,

$$\lim_{n\to\infty} \frac{1}{n} \sum_{i=0}^{n-1} \mathbb{I}_I(R^i(x)) = \lambda(I). \tag{5.22}$$

As the collection of dyadic intervals is a countable collection of sets, there exists a point $x \in [0,1)$ (in fact, there is a full measure set of such points) so that (5.22) holds for all dyadic intervals J. Now let I be an arbitrary interval. For any $\varepsilon > 0$ there exist dyadic intervals J and K such that $J \subset I \subset K$ and $\lambda(K \setminus J) < \varepsilon$. From the nature of the intervals, it follows that

$$\frac{1}{n}\sum_{i=0}^{n-1} \mathbb{I}_J(R^i(x)) \le \frac{1}{n}\sum_{i=0}^{n-1} \mathbb{I}_I(R^i(x)) \le \frac{1}{n}\sum_{i=0}^{n-1} \mathbb{I}_K(R^i(x)).$$

Then

$$\lambda(J) = \lim_{n\to\infty} \frac{1}{n} \sum_{i=0}^{n-1} \mathbb{I}_J(R^i(x))$$

$$\leq \liminf_{n\to\infty} \frac{1}{n} \sum_{i=0}^{n-1} \mathbb{I}_I(R^i(x))$$

$$\leq \lim_{n\to\infty} \frac{1}{n} \sum_{i=0}^{n-1} \mathbb{I}_K(R^i(x)) = \lambda(J) + \varepsilon.$$

Similarly,

$$\lambda(J) \leq \limsup_{n\to\infty} \frac{1}{n} \sum_{i=0}^{n-1} \mathbb{I}_I(R^i(x)) \leq \lambda(J) + \varepsilon.$$

As this holds for all ε and $\lambda(J) < \lambda(I) < \lambda(J) + \varepsilon$, then

$$\liminf_{n\to\infty} \frac{1}{n} \sum_{i=0}^{n-1} \mathbb{I}_I(R^i(x)) = \limsup_{n\to\infty} \frac{1}{n} \sum_{i=0}^{n-1} \mathbb{I}_I(R^i(x)).$$

From this it follows that for the point x,

$$\lim_{n\to\infty} \frac{1}{n} \sum_{i=0}^{n-1} \mathbb{I}_I(R^i(x)) = \lambda(I)$$

for all intervals I. Finally we observe that given any interval I there is an interval I' of the same measure so that $R^i(0) \in I$ if and only if $R^i(x) \in I'$. This completes the proof of the theorem. $\square$

Exercises

(1) Show that the digit $d, 0 < d < 10$, occurs as a first digit in powers of 2 with frequency $\log_{10}(d+1)/d$.

(2) Let ν be a probability measure that is ergodic and invariant for an irrational rotation R on $[0,1)$. Show that ν must be Lebesgue measure.

5.4. The Mean Ergodic Theorem

The Mean Ergodic Theorem was proved by von Neumann in 1931. It asserts the convergence of the ergodic averages in the norm of L^2. Later it was shown that convergence in the norm of L^p also holds. In

5.4. The Mean Ergodic Theorem

general this is a weaker result than almost everywhere convergence, but it is useful in many cases. In particular, this theorem is sufficient to prove the characterization of ergodicity in Theorem 5.1.5. We will present two proofs: one that uses the special properties of L^2, and the other that derives convergence in L^1 as a consequence of the Birkhoff ergodic theorem.

Throughout this section, $(X, \mathcal{S}, \mu)$ will be a probability space. Let $\infty > p \geq 1$, and let $||\cdot||_p$ denote the norm in L^p, i.e., $||f||_p = (\int |f|^p d\mu)^{\frac{1}{p}}$. We say that a sequence of functions $\{f_n\}$ in $L^p(X, \mu)$ converges to a function $f \in L^p$ in p-**norm** or L^p-**norm** or in **mean** if

$$\lim_{n \to \infty} ||f_n - f||_p = 0.$$

We first present two standard examples showing that, in general, convergence in L^p and a.e. convergence (as in the Birkhoff ergodic theorem) are independent notions. For both examples the measure space is $[0, 1]$ with Lebesgue measure. First define a family of functions f_n by

$$f_n = 2^n \mathbb{I}_{[0, 1/n]}.$$

Observe that for all $x \in (0, 1]$, $f_n(x) \to 0$ as $n \to \infty$. So f_n converges to 0 a.e. Now, we compute

$$||f_n - 0||_p^p = ||f_n||_p^p = \frac{2^{pn}}{n}.$$

So, $||f_n||_p \to \infty$ as $n \to \infty$, showing that a.e. convergence does not imply L^p convergence. We note, however, that there exist conditions under which a.e. convergence implies L^p convergence. One such condition is discussed in Exercise 2; a more general condition under which this holds is called uniform integrability, for which the reader may refer to the Bibliographical Notes.

For our second example, we define a family of functions that take the values of 0 and 1 on various subintervals of $[0, 1]$. Recall that a dyadic interval of order n in $[0, 1)$ has the form $[\frac{k}{2^n}, \frac{k+1}{2^n})$ for $k = 0, \ldots, 2^n - 1$. List all the dyadic intervals in the following order: first list those of order 1 from left to right, then those of order 2 from left to right, etc. This gives a countable collection of dyadic intervals that we rename $\{K_i\}_{i \geq 1}$. Let $g_i = \mathbb{I}_{K_i}$. We claim that the sequence $\{g_i\}$ does not converge a.e. This is a consequence of the fact that for each $x \in [0,1)$ there are infinitely many values of i such that $g_i(x) = 0$, and there are infinitely many values of i such that $g_i(x) = 1$. This implies that $\{g_i\}$ does not converge a.e. Now to compute the L^p norm of functions in this sequence assume that K_i is a dyadic interval of order n. Then $\lambda(K_i) = \frac{1}{2^n}$, so

$$\|g_i\|_p^p = \lambda(K_i) = \frac{1}{2^n}.$$

As i increases, the order of K_i increases, showing that the sequence $\|g_i\|_p$ converges to 0, so g_i converges to 0 in L^p. This shows that L^p convergence does not imply a.e. convergence.

Before we start with the proof of the L^2 ergodic theorem, we need to introduce some notions that exploit the rich structure of L^2. In particular, we use that L^2 has an inner product. The inner product is closely related to the L^2 norm. As we have stated, L^2 with this norm (after functions are identified a.e.) is complete as a metric space. A vector space with an inner product that is complete is called a *Hilbert space*, and many of the properties we shall discuss for L^2 extend in a natural way to Hilbert spaces.

The first notion we need to introduce is that of the orthogonal complement of a (closed linear) subspace of L^2. Here we will appeal to the reader's intuition that comes from studying this notion in linear algebra. When working in $\mathbb{R}^2$, say, with its inner product, given a subspace S consisting of a line through the origin, there is defined a unique line through the origin orthogonal to it, its *orthogonal complement* denoted by $S^\perp$. Also, any point in $\mathbb{R}^2$ can be written in a unique way as the sum of a point in S with a point in $S^\perp$; in this case one says that $\mathbb{R}^2$ is the *direct sum* of S and $S^\perp$. We state Lemma 5.4.1 with only a brief outline; a complete proof may be found in [**67**, Lemma 4.1]. It gives the existence of orthogonal complements of closed subspaces of L^2 (i.e., S is itself a vector space and it is closed in the metric given by the L^2-norm). The **orthogonal complement** of S, denoted $S^\perp$, is defined by

$$S^\perp = \{g \in S : (f,g) = 0 \text{ for all } f \in S\}.$$

5.4. The Mean Ergodic Theorem

Using the orthogonal complement of a closed subspace S, one can define a **projection** $P : L^2 \to S$ so that $P(f) = g$ for any $f \in L^2$ where g is the unique element of S so that $f = g + h$ for $h \in S^\perp$.

Lemma 5.4.1. *Let S be a closed linear subspace of L^2. Then for each element f in L^2 there exists a unique element f_0 in S that is closest to f in the sense that*

$$\inf\{||f - g||_2 : g \in S\} = ||f - f_0||_2.$$

*Furthermore, if $P : L^2 \to S$ is defined by $P(f) = f_0$, then P is a projection, called the **projection onto** S, and every element f of L^2 can be written in a unique way as*

$$f = P(f) + h,$$

where h is in the orthogonal complement of S.

Proof. We give an outline of the proof so that the reader knows the techniques that are used. First note that when f is in S, all we need to do is to define f_0 to be f. When $f \notin S$, then as S is closed $\inf\{||f-g||_2 : g \in L^2\} > 0$. Choose a sequence of functions g_n so that $||f - g_n||_2$ converges to $\inf\{||f - g||_2 : g \in S\}$. Then there is some technical work to show that the sequence $\{g_n\}$ is a Cauchy sequence. Therefore it must converge as the space L^2 is complete. Then one lets f_0 be the limit of this sequence. Next one has to show that every $g \in S$ is orthogonal to $f - f_0$. This can also be used to show the uniqueness of f_0. We are ready to set $P(f) = f_0$. Finally, write f as $f = P(f) + f - P(f)$. It can be shown from the construction of $P(f)$ that $f - P(f)$ is in the orthogonal complement of S. □

Theorem 5.4.2 (Mean Ergodic Theorem). *Let $(X, \mathcal{S}, \mu)$ be a probability space and let $T : X \to X$ be a measure-preserving transformation. Let $\mathcal{I} = \{g \in L^2 : g \circ T = g \text{ a.e.}\}$ be the subspace of T-invariant functions and let P be the projection from L^2 to $\mathcal{I}$. Then for any $f \in L^2$, the sequence*

(5.23) $$\frac{1}{n} \sum_{i=0}^{n-1} f \circ T^i$$

converges in L^2 to $P(f)$.

Proof. The structure of the proof is to decompose L^2 into two closed subspaces, to verify that the theorem holds on each subspace (each case holding different reasons) and finally to obtain the proof by putting together the two special cases.

When $f \in \mathcal{I}$, $f \circ T^n = f$ a.e., so

$$\frac{1}{n}\sum_{n=0}^{n-1} f \circ T^i = f = P(f).$$

Thus, the limit in (5.23) clearly exists and converges to $P(f)$.

Next, we consider a subspace B of functions,

$$B = \{f : f = g \circ T - g, g \in L^2\}.$$

(The elements of B are called (additive) co-boundaries.) Observe that if $f = g \circ T - g$, then

$$\sum_{i=0}^{n-1} f \circ T^i = \sum_{i=0}^{n-1} g \circ T^{i+1} - g \circ T^i$$
$$= g \circ T^n - g.$$

Next we show that the limit in (5.23) also converges to 0 for functions in the norm closure of B, denoted $\overline{B}$. So let f_j be a sequence in B converging to a function f in L^2. Then

$$\left\|\frac{1}{n}\sum_{i=0}^{n-1} f \circ T^i\right\|_2 = \left\|\frac{1}{n}\sum_{i=0}^{n-1}(f - f_j + f_j) \circ T^i\right\|_2$$
$$\leq \frac{1}{n}\sum_{i=0}^{n-1}\|(f - f_j) \circ T^i\|_2 + \frac{1}{n}\left\|\sum_{i=0}^{n-1} f_j \circ T^i\right\|_2$$
$$= \frac{1}{n}\sum_{i=0}^{n-1}\|(f - f_j)\|_2 + \frac{1}{n}\left\|\sum_{i=0}^{n-1} f_j \circ T^i\right\|_2.$$

Therefore the limit above converges to 0.

Now we show that the orthogonal complement of the closure of B is $\mathcal{I}$. First let $f \in \mathcal{I}$. Write $h \in B$ as $h = g \circ T - g$ for $g \in L^2$.

5.4. The Mean Ergodic Theorem

Then
$$(f, h) = (f, g \circ T - g) = (f, g \circ T) - (f, g)$$
$$= (f \circ T, g \circ T) - (f, g)$$
$$= (f, g) - (f, g) = 0.$$

Now let $h \in \overline{B}$. Let h_k be a sequence in B converging to h in norm. We know that
$$\lim_{k \to \infty} (h - h_k, f) = 0,$$
and $(f, h_k) = 0$ for all $k > 0$. So $(f, h) = 0$, showing that $\mathcal{I}$ is contained in $\overline{B}^\perp$.

Next let $f \in \overline{B}^\perp$. Then for every $g \in L^2$, $(f, g \circ T - g) = 0$. So,
$$(f, g) = (f, g \circ T) = \int f \bar{g} \circ T \, d\mu$$
$$= \int f \circ T^{-1} \bar{g} \, d\mu = (f \circ T^{-1}, g).$$

As $(f - f \circ T, g) = 0$ for all $g \in L^2$, $f = f \circ T$ a.e. This completes the proof that $\mathcal{I} = \overline{B}^\perp$. Now let P be the projection onto the subspace $\mathcal{I}$. By Lemma 5.4.1, every element $f \in L^2$ can be written in a unique way as $f = P(f) + h$, where $P(f) \in \mathcal{I}$ and $h \in \overline{B}$. We have already seen that the ergodic theorem holds for g and h, so it holds for f and the limit converges in L^2 norm to $P(f)$. □

We now present a different proof for the mean ergodic theorem in L^1. We will obtain it as a consequence of the following lemma.

Lemma 5.4.3 (Scheffé's Lemma). *Let $f_n, n \geq 1$, and f be nonnegative functions in L^1 such that $f_n \to f$ a.e. Then f_n converges to f in L^1 if and only if $\int f_n \, d\mu$ converges to $\int f \, d\mu$.*

Proof. We show the harder part: if $\int f_n \, d\mu \to \int f \, d\mu$, then f_n converges to f in L^1.

As f_n and f are nonnegative, it follows that $(f_n - f)^- \leq f$. Then Lebesgue's dominated convergence theorem implies that
$$\lim_{n \to \infty} \int (f_n - f)^- \, d\mu = \int \lim_{n \to \infty} (f_n - f)^- = 0.$$

Next let $A_n = \{x \in X : f_n(x) \geq f(x)\}$. Then

$$\int (f_n - f)^+ \, d\mu = \int_{A_n} f_n - f \, d\mu$$
$$= \int_X f_n - f \, d\mu - \int_{A_n^c} f_n - f \, d\mu.$$

Now on A_n^c, $(f_n - f)^- = -(f_n - f)$. So,

$$\lim_{n \to \infty} \left| \int_{A_n^c} f_n - f \, d\mu \right| \leq \lim_{n \to \infty} \int_X (f_n - f)^- \, d\mu = 0.$$

Therefore, $\lim_{n \to \infty} \int (f_n - f)^+ \, d\mu = 0$, completing the proof. □

We show that the theorem follows from the Birkhoff ergodic theorem.

Theorem 5.4.4 (Mean Ergodic Theorem in L^1). *Let $(X, \mathcal{S}, \mu)$ be a probability space and let $T : X \to X$ be a measure-preserving transformation. Then for any $f \in L^1$, the sequence*

(5.24) $$f_n(x) = \frac{1}{n} \sum_{i=0}^{n-1} f \circ T^i$$

converges in L^1 norm to a function $\tilde{f} \in L^1$.

Proof. The reader should verify that it suffices to prove the theorem when f is nonnegative. Let $f_n(x) = \sum_{i=0}^{n-1} f(T^i(x))$. Then the functions $(1/n)f_n$ are nonnegative and by the Birkhoff ergodic theorem converge a.e. to a function $\tilde{f} \in L^1$. Also,

$$\int (1/n) f_n \, d\mu = \frac{1}{n} \sum_{i=0}^{n-1} \int f \circ T^i \, d\mu = \frac{1}{n} \sum_{i=0}^{n-1} \int f \, d\mu = \int f \, d\mu.$$

Then Scheffé's lemma implies that $(1/n)f_n$ converges to $\tilde{f}$ in L^1. □

We conclude with a statement of the mean ergodic theorem in L^p.

5.4. The Mean Ergodic Theorem

Theorem 5.4.5 (Mean Ergodic Theorem in L^p). *Let $(X, \mathcal{S}, \mu)$ be a probability space and let $T : X \to X$ be a measure-preserving transformation. Then for any $f \in L^p$, the sequence*

(5.25) $$f_n(x) = \frac{1}{n} \sum_{i=0}^{n-1} f \circ T^i$$

converges in L^p norm to a function $\tilde{f} \in L^p$.

Exercises

(1) Let $P : L^2 \to S$ be the projection operator. Show that P is a linear transformation.

(2) Prove the following extension of the dominated convergence theorem. Let $(X, \mathcal{S}, \mu)$ be a measure space, and let $f_n : X \to \mathbb{R}$ be a sequence of measurable functions converging to $f(x)$. Suppose that for $p \geq 1$ there exists a function $g \in L^1$ such that $|f_n(x)| \leq g(x)$ a.e. for all $n \geq 1$. Then f is in L^p and $||f_n - f||_p \to 0$ as $n \to \infty$.

* (3) Prove Theorem 5.4.5.

(4) Let $f_n, n \geq 1$, and f be nonnegative functions in L^1 such that $f_n \to f$ a.e. Show that $\lim_{n \to \infty} \int |f_n - f| \, d\mu = 0$ if and only if $\int |f_n| \, d\mu$ converges to $\int |f| \, d\mu$.

Chapter 6

Mixing Notions

6.1. Introduction

This chapter introduces several notions of mixing for transformations that are measure-preserving on a probability space $(X, \mathcal{S}, \mu)$. For concreteness the reader may think of X as the unit interval with Lebesgue measure.

We saw in Theorem 5.1.5 that a measure-preserving transformation T on a probability space $(X, \mathcal{S}, \mu)$ is ergodic if and only if for all measurable sets A and B,

$$(6.1) \qquad \lim_{n \to \infty} \frac{1}{n} \sum_{i=0}^{n-1} \mu(T^{-i}(A) \cap B) = \mu(A)\mu(B).$$

A measure-preserving transformation T on a probability space $(X, \mathcal{S}, \mu)$ is said to be **mixing** if for all measurable sets A, B,

$$(6.2) \qquad \lim_{n \to \infty} \mu(T^{-n}(A) \cap B) = \mu(A)\mu(B).$$

We can also write the mixing condition, provided that $\mu(B) > 0$, as

$$\lim_{n \to \infty} \frac{\mu(T^{-n}(A) \cap B)}{\mu(B)} = \mu(A),$$

which may be interpreted as saying that "after some time," the proportion of A that is in B (under iteration by T) is close to the size of

A (which is the proportion of A in X). Also, we see from (6.1) that ergodicity can be thought of as "mixing on the average."

In the early literature, mixing was called strong mixing, to distinguish it from weak mixing. Mixing can also be seen as a statement about asymptotic independence. Recall that in probability theory, two *events*, represented by measurable sets A and B, are said to be *independent* if $\mu(A \cap B) = \mu(A)\mu(B)$. If T is mixing $T^{-n}(A)$ is "eventually independent" from B.

For any measurable sets A and B write

$$a_i = a_i(A, B) = \mu(T^{-i}(A) \cap B).$$

Mixing can then be expressed as the convergence of the sequences

$$a_i(A, B) \to \mu(A)\mu(B) \text{ as } i \to \infty$$

for all measurable sets A and B. Similarly, ergodicity is equivalent to the statement that

$$\frac{1}{n}\sum_{i=0}^{n-1} a_i(A, B) \to \mu(A)\mu(B) \text{ as } n \to \infty$$

for all measurable sets A and B. We now explore three different ways in which a bounded sequence $\{a_i\}$ may converge to a number a.

(1) Convergence of sequences:
$$\lim_{i \to \infty} a_i - a = 0;$$

(2) Strong Cesàro convergence of sequences:
$$\lim_{n \to \infty} \frac{1}{n}\sum_{i=0}^{n-1} |a_i - a| = 0;$$

(3) Cesàro convergence of sequences:
$$\lim_{n \to \infty} \frac{1}{n}\sum_{i=0}^{n-1} a_i - a = 0.$$

There are some basic implications among these three different convergence notions.

Lemma 6.1.1. *Let $\{a_i\}$ be a bounded sequence. Then convergence implies strong Cesàro convergence, and strong Cesàro convergence*

6.1. Introduction

implies Cesàro convergence. Furthermore, the converse implications do not hold.

Proof. To show that (1) implies (2), let $\varepsilon > 0$. Then there exists $N_1 > 0$ such that for $i \geq N_1$, $|a_i - a| < \varepsilon/2$. As $\{a_i\}$ is bounded there exists $N_2 > 0$ such that for $n > N_2$

$$\frac{1}{n}\sum_{i=0}^{N_1-1}|a_i - a| < \frac{\varepsilon}{2}.$$

Let $N = \max\{N_1, N_2\}$. Then if $n > N$,

$$\frac{1}{n}\sum_{i=0}^{n-1}|a_i - a| = \frac{1}{n}\sum_{i=0}^{N_1-1}|a_i - a| + \frac{1}{n}\sum_{i=N_1}^{n-1}|a_i - a|$$
$$= \frac{\varepsilon}{2} + \frac{1}{n}(n - N_1) \cdot \frac{\varepsilon}{2}$$
$$< \frac{\varepsilon}{2} + \frac{\varepsilon}{2} = \varepsilon.$$

The fact that (2) implies (3) is a direct consequence of the triangle inequality. The proof that the implications are proper is left to the reader. $\square$

To study another characterization of Cesàro convergence we need the notion of zero density. A set D of nonnegative integers is said to be of **zero density** if

$$\lim_{n\to\infty}\frac{1}{n}\sum_{i=0}^{n-1}\mathbb{I}_D(i) = 0.$$

Otherwise (i.e., if $\limsup_{n\to\infty}\frac{1}{n}\sum_{i=0}^{n-1}\mathbb{I}_D(i) > 0$), it is said to be of **positive density**. A sequence of nonnegative integers $\{a_i\}_{i\geq 0}$ is said to be of zero density if the set $\{a_i : i \geq 0\}$ is of zero density. Clearly, any finite set has density zero, but there are infinite sets of zero density. For example, the sequence of powers of 2 ($a_i = 2^i$) is a zero density sequence (Exercise 1), while the sequence of even integers is a positive density sequence. It also follows from the Prime Number Theorem that the sequence of primes is a density zero sequence. A set $A \subset \{0, 1, \ldots\}$ is said to be a **density one** set if its complement has zero density. The following proposition characterizes Cesàro convergence for bounded *nonnegative* sequences. (Note that for nonnegative sequences, strong Cesàro convergence to 0 and Cesàro convergence to 0 are equivalent.)

We first define a new notion of convergence. A sequence $\{a_i\}$ of real numbers is said to **converge in density** to a point a if there exists a zero density set $D \subset \mathbb{N}$ such that for all $\varepsilon > 0$ there is an integer N such that for all $i > N, i \notin D$, $|a_i - a| < \varepsilon$. In this case we write
$$D - \lim_{i \to \infty} a_i = a \quad \text{or} \quad \lim_{i \to \infty, i \notin D} a_i = a.$$

Proposition 6.1.2. *Let $\{b_i\}$ be a bounded sequence of nonnegative real numbers. Then the sequence $\{b_i\}$ converges Cesàro to 0 if and only if $\{b_i\}$ converges in density to 0.*

Proof. Let the sequence $\{b_i\}$ be bounded by β. Suppose that b_i converges to 0 outside a zero density set D: $\lim_{i \to \infty, i \notin D} b_i = 0$. Then

$$\lim_{n \to \infty} \frac{1}{n} \sum_{i=0}^{n-1} b_i = \lim_{n \to \infty} \frac{1}{n} \left[\sum_{i=0, i \in D}^{n-1} b_i + \sum_{i=0, i \notin D}^{n-1} b_i \right]$$
$$\leq \lim_{n \to \infty} \frac{1}{n} \sum_{i=0, i \in D}^{n-1} \beta + \lim_{n \to \infty} \frac{1}{n} \sum_{i=0, i \notin D}^{n-1} b_i$$
$$= \beta \lim_{n \to \infty} \frac{1}{n} \sum_{i=0}^{n-1} \mathbb{I}_D(i) + \lim_{n \to \infty} \frac{1}{n} \sum_{i=0, i \notin D}^{n-1} b_i = 0.$$

Conversely, assume that
$$\lim_{n \to \infty} \frac{1}{n} \sum_{i=0}^{n-1} b_i = 0.$$

To construct the zero density set, first define for each $k > 0$ the sets
$$D_k = \{i : \frac{1}{k} < b_i\}.$$

Then
$$\frac{1}{k} \mathbb{I}_{D_k}(i) < b_i.$$

6.2. Weak Mixing

Evidently $D_k \subset D_{k+1}$, and since

$$\frac{1}{k}\left(\lim_{n\to\infty}\frac{1}{n}\sum_{i=0}^{n-1}\mathbb{I}_{D_k}(i)\right) = \lim_{n\to\infty}\frac{1}{n}\sum_{i=0}^{n-1}\frac{1}{k}\mathbb{I}_{D_k}(i) \leq \lim_{n\to\infty}\frac{1}{n}\sum_{i=0}^{n-1}b_i = 0,$$

each D_k has zero density. In particular, this implies that for each $k > 0$ we may choose $n_k > n_{k-1} > 0$ (set $n_0 = 0$) so that

$$\frac{1}{n}\sum_{i=0}^{n-1}\mathbb{I}_{D_k}(i) < \frac{1}{k}$$

for all $n > n_k$. Then define D by

$$D = \bigsqcup_{k>0} D_k \cap [n_k, n_{k+1}).$$

It is clear that b_i converges to 0 outside D. The proof that D has zero density is left to the reader. □

Exercises

(1) Show that the sequence of powers of 2 is a zero density sequence.

(2) Show that the union of finitely many zero density sets has zero density.

(3) Show that the converses of the implications in Lemma 6.1.1 do not hold.

(4) Give the details in the proof of Proposition 6.1.2 to show that b_i converges to 0 outside the set D.

* (5) Show that the set D in Proposition 6.1.2 has zero density.

(6) Show that Proposition 6.1.2 fails when b_i is not nonnegative.

6.2. Weak Mixing

A useful property that lies between ergodicity and mixing is weak mixing. A measure-preserving transformation T on a probability space $(X, \mathcal{S}, \mu)$ is said to be **weakly mixing** if

(6.3) $$\lim_{n\to\infty}\frac{1}{n}\sum_{i=0}^{n-1}|\mu(T^{-i}(A)\cap B) - \mu(A)\mu(B)| = 0.$$

If we let $a_i = a_i(A, B) = \mu(T^{-i}(A) \cap B)$, then it is clear that weak mixing is equivalent to the requirement that for each pair of measurable sets A and B, the bounded sequence a_i converges strong Cesàro to $\mu(A)\mu(B)$.

Lemma 6.2.1. *Let T be a probability-preserving transformation.*

(1) *If T is weakly mixing, then it is ergodic.*

(2) *If T is mixing, then it is weakly mixing.*

Proof. If T is weakly mixing, then the sequence

$$a_i(A, B) = \mu(T^{-i}(A) \cap B)$$

converges strong Cesàro to $\mu(A)\mu(B)$, so the sequence $a_i(A, B)$ converges Cesàro to $\mu(A)\mu(B)$. Therefore T is ergodic. (There is also a direct proof: if A is T-invariant, then $a_i(A, A^c) = 0$, but since it converges to $\mu(A)\mu(A^c)$ it follows that $\mu(A) = 0$ or $\mu(A^c) = 0$.) If T is mixing, then the sequence $a_i(A, B)$ converges to $\mu(A)\mu(B)$, so it converges strong Cesàro to $\mu(A)\mu(B)$. Therefore T is weakly mixing. □

The fact that the converses of these two implications are not true does not necessarily follow from the analogous result for bounded sequences as there is no *a priori* reason to think that the counterexamples for bounded sequences can be realized by transformations. We prove in this section that ergodicity does not imply weak mixing; that weak mixing does not imply mixing will be left for later after we construct a weakly mixing transformation that is not mixing. First we study other conditions equivalent to weak mixing. Our first result is a direct consequence of the properties of bounded sequences.

Proposition 6.2.2. *Let T be a measure-preserving transformation on the probability space $(X, \mathcal{S}, \mu)$. Then the following are equivalent.*

(1) *T is weakly mixing.*

(2) *For each pair of measurable sets A and B there is a zero density set $D = D(A, B)$ so that*

$$\lim_{i \to \infty, i \notin D} \mu(T^{-i}(A) \cap B) = \mu(A)\mu(B).$$

6.2. Weak Mixing

(3) *For each pair of measurable sets A and B*

$$\lim_{n \to \infty} \frac{1}{n} \sum_{i=0}^{n-1} \left(\mu(T^{-i}(A) \cap B) - \mu(A)\mu(B)\right)^2 = 0.$$

Proof. The fact that (1) and (2) are equivalent follows directly from Proposition 6.1.2 applied to the nonnegative bounded sequence

$$b_i = |\mu(T^{-i}(A) \cap B) - \mu(A)\mu(B)|.$$

Now suppose that (1) holds. Then clearly

$$\lim_{i \to \infty, i \notin D} \left(\mu(T^{-i}(A) \cap B) - \mu(A)\mu(B)\right)^2 = 0,$$

and by Proposition 6.1.2

$$\lim_{n \to \infty} \frac{1}{n} \sum_{i=0}^{n-1} \left(\mu(T^{-i}(A) \cap B) - \mu(A)\mu(B)\right)^2 = 0.$$

The converse is similar. □

We note that part (2) of Proposition 6.2.2 can be made stronger (see Lemma 6.3.5).

We conclude with a property that is equivalent to weak mixing. A measure-preserving transformation T on a measure space $(X, \mathcal{S}, \mu)$ is said to be **doubly ergodic** if for all measurable sets A and B of positive measure there exists an integer $n > 0$ such that

$$\mu(T^{-n}(A) \cap A) > 0 \quad \text{and} \quad \mu(T^{-n}(A) \cap B) > 0.$$

Note that the "time" n at which the image of A under T^{-n} intersects A and B is required to be the same. It is clear that if T is doubly ergodic, then it is recurrent and ergodic. Later we will show the equivalence of double ergodicity and several other characterizations of weak mixing, but for now we only prove one direction of the equivalence.

Proposition 6.2.3. *Let T be a measure-preserving transformation on the probability space $(X, \mathcal{S}, \mu)$. If T is weakly mixing, then T is doubly ergodic.*

Proof. Suppose that T is weakly mixing. Let A and B be measurable sets of positive measure. Then there there exist zero density sets $D_1 = D(A,B)$ and $D_2 = D(A,A)$ such that

$$\lim_{i \to \infty, i \notin D_1} \mu(T^{-i}(A) \cap B) = \mu(A)\mu(B) \text{ and}$$

$$\lim_{i \to \infty, i \notin D_2} \mu(T^{-i}(A) \cap A) = \mu(A)\mu(A).$$

It follows that there is an integer $i > 0$ such that $\mu(T^{-i}(A) \cap A) > 0$ and $\mu(T^{-i}(A) \cap B) > 0$. Thus T is doubly ergodic. The converse is proved in Section 6.4. □

Lemma 6.2.4. *Rotations are not weakly mixing.*

Proof. Let $I = [0, 1/8)$ and $J = [1/2, 5/8)$ be two intervals in $[0,1)$ (any other small intervals that are sufficiently apart will work). Then for any integer n such that $T^n(I) \cap J \neq \emptyset$ it follows that $T^n(I) \cap I = \emptyset$. Therefore R is not doubly ergodic, so not weakly mixing. □

Exercises

(1) Show that the binary odometer is not weakly mixing.

(2) Let T be a measure-preserving transformation. Show that if T is doubly ergodic, then so is T^2. What about T^k for $k > 0$?

(3) Show that if T is the dyadic odometer and S is any finite measure-preserving transformation, then $T \times S$ is not weakly mixing.

(4) Show that if T is weakly mixing, then any factor of T is weakly mixing.

(5) Let T be a measure-preserving transformation. Show that T is doubly ergodic if and only if for all pairs of measurable sets A and B of positive measure there exists an integer $n > 0$ so that $T^{-n}(A) \cap A \neq \emptyset$ and $T^{-n}(A) \cap B \neq \emptyset$.

(6) Observe that the double ergodicity definition also holds for infinite measure and show that the Hajian-Kakutani transformation is not doubly ergodic.

6.3. Approximation

This section presents some approximation theorems that are used to show that for some dynamical properties (such as mixing) it is sufficient to verify the property for each element of a sufficient semi-ring.

We have seen that if $\mathcal{C}$ is a sufficient semi-ring for a probability space $(X, \mathcal{S}, \mu)$ and A is a set of positive measure in X, then for $\varepsilon > 0$ there exist $I_1, \ldots, I_p$ in $\mathcal{C}$ such that

$$\mu(A \triangle (\bigcup_{j=1}^{p} I_j)) < \varepsilon.$$

So finite unions of elements of a sufficient semi-ring approximate measurable sets "up to ε". Recall that the collection of all finite unions of elements of a semi-ring is a ring (see Exercise 2.7.3). A **sufficient ring** is a ring generated by a sufficient semi-ring. For our approximation results we have to discuss two steps, one lifting the property from a semi-ring to the ring it generates, and the next lifting the property from the sufficient ring to the measurable sets.

The following is a technical lemma. It will be used when A and B are arbitrary measurable sets and E and F come from a sufficient semi-ring or ring.

Lemma 6.3.1. *Let $(X, \mathcal{S}, \mu)$ be a measure space and T a measure-preserving transformation. Then for any integer $n > 0$ and any measurable sets A, B, E, F,*

$$\mu([T^{-n}(A) \cap B] \triangle [T^{-n}(E) \cap F]) \leq \mu(A \triangle E) + \mu(B \triangle F).$$

Proof. By the triangle inequality (see Exercise 2.5.6),

$$\mu([T^{-n}(A) \cap B] \triangle [T^{-n}(E) \cap F])$$
$$\leq \mu([T^{-n}(A) \cap B] \triangle [T^{-n}(A) \cap F])$$
$$+ \mu([T^{-n}(A) \cap F] \triangle [T^{-n}(E) \cap F]).$$

As $(A \cap E) \triangle (B \cap E) = (A \triangle B) \cap E$,

$$\mu([T^{-n}(A) \cap B] \triangle [T^{-n}(E) \cap F])$$
$$\leq \mu(T^{-n}(A) \cap (B \triangle F)) + \mu(T^{-n}(A \triangle E) \cap F)$$
$$\leq \mu(B \triangle F) + \mu(T^{-n}(A \triangle E))$$
$$= \mu(B \triangle F) + \mu(A \triangle E).$$

□

The following lemma provides the main tool used in the proofs of the approximation results.

Lemma 6.3.2. *Let T be a finite measure-preserving transformation on a probability measure space $(X, \mathcal{S}, \mu)$. Suppose that for $A, B \in \mathcal{S}$ and $\varepsilon > 0$ there exist sets $E, F \in \mathcal{S}$ such that*

$$\mu(A \triangle E) < \varepsilon \text{ and } \mu(B \triangle F) < \varepsilon.$$

If for $n > 0$

$$|\mu(T^{-n}(E) \cap F) - \mu(E)\mu(F)| < \varepsilon,$$

then

$$|\mu(T^{-n}(A) \cap B) - \mu(A)\mu(B)| < 5\varepsilon.$$

Proof. By Lemma 6.3.1,

$$\mu([T^{-n}(A) \cap B] \triangle [T^{-n}(E) \cap F]) \leq \mu(A \triangle E) + \mu(B \triangle F) < 2\varepsilon.$$

Using Exercise 2.5.5, this inequality yields

$$|\mu(T^{-n}(A) \cap B) - \mu(T^{-n}(E) \cap F)| < 2\varepsilon.$$

Then

$$|\mu(T^{-n}(A) \cap B) - \mu(A)\mu(B)| \leq |\mu(T^{-n}(A) \cap B) - \mu(T^{-n}(E) \cap F)|$$
$$+ |\mu(T^{-n}(E) \cap F) - \mu(E)\mu(F)|$$
$$+ |\mu(E)\mu(F) - \mu(A)\mu(F)|$$
$$+ |\mu(A)\mu(F) - \mu(A)\mu(B)|$$
$$< 2\varepsilon + \varepsilon + \varepsilon\mu(F) + \varepsilon\mu(A)$$
$$\leq 5\varepsilon.$$

□

6.3. Approximation

The following lemma has a similar proof.

Lemma 6.3.3. *Let T be a finite measure-preserving transformation on a probability space $(X, \mathcal{S}, \mu)$. Suppose that for $A, B \in \mathcal{S}$ and $\varepsilon > 0$ there exist sets $E, F \in \mathcal{S}$ such that*

$$\mu(A \triangle E) < \varepsilon \text{ and } \mu(B \triangle F) < \varepsilon.$$

(1) *If for $n > 0$*

$$\left| \frac{1}{n} \sum_{i=0}^{n-1} \mu(T^{-i}(E) \cap F) - \mu(E)\mu(F) \right| < \varepsilon,$$

then

$$\left| \frac{1}{n} \sum_{i=0}^{n-1} \mu(T^{-i}(A) \cap B) - \mu(A)\mu(B) \right| < 5\varepsilon.$$

(2) *If for $n > 0$*

$$\frac{1}{n} \sum_{i=0}^{n-1} \left| \mu(T^{-i}(E) \cap F) - \mu(E)\mu(F) \right| < \varepsilon,$$

then

$$\frac{1}{n} \sum_{i=0}^{n-1} \left| \mu(T^{-i}(A) \cap B) - \mu(A)\mu(B) \right| < 5\varepsilon.$$

Theorem 6.3.4 shows that to prove the limit definition of ergodicity, weak mixing, or mixing it suffices to verify it for all elements of a sufficient semi-ring. Note, however, that as shown in Section 6.8, the condition: for all sets I and J of positive measure in a sufficient semi-ring, there exists an integer $n > 0$ such that $\mu(T^{-n}(I) \cap J) > 0$, does not imply that T is ergodic. Therefore, in the case of ergodicity (and weak mixing) there are some characterizations that need to be verified for all measurable sets and not just for elements of a sufficient semi-ring, while for other equivalent characterizations, such as those in Theorem 6.3.4, it is enough to verify them on a sufficient semi-ring.

Theorem 6.3.4. *Let T be a measure-preserving transformation on a probability space $(X, \mathcal{S}, \mu)$ with a sufficient semi-ring $\mathcal{C}$.*

(1) If for all $I, J \in \mathcal{C}$,
$$\lim_{n\to\infty} \frac{1}{n} \sum_{i=0}^{n-1} \mu(T^{-i}(I) \cap J) = \mu(I)\mu(J),$$
then T is ergodic.

(2) If for all $I, J \in \mathcal{C}$,
$$\lim_{n\to\infty} \frac{1}{n} \sum_{i=0}^{n-1} \left|\mu(T^{-i}(I) \cap J) - \mu(I)\mu(J)\right| = 0,$$
then T is weakly mixing.

(3) If for all $I, J \in \mathcal{C}$,
$$\lim_{n\to\infty} \mu(T^{-n}(I) \cap J) = \mu(I)\mu(J),$$
then T is mixing.

Proof. We prove part (3); the proofs of the other parts are similar and left to the reader. Let A and B be sets of positive measure. By Lemma 2.7.3, for any $\varepsilon > 0$ there exist $I_j, J_k \in \mathcal{C}, j = 1, \ldots, p, k = 1, \ldots, q$, such that if we write $E = \bigsqcup_{j=1}^{p} I_j$ and $F = \bigsqcup_{k=1}^{q} J_k$, then
$$\mu(A \triangle E) < \varepsilon \text{ and } \mu(B \triangle F) < \varepsilon.$$

For each $k \in \{1, \ldots, q\}$,
$$\lim_{n\to\infty} \mu(T^{-n}(E) \cap J_k) = \lim_{n\to\infty} \mu(T^{-n}(\bigsqcup_{j=1}^{p} I_j) \cap J_k)$$
$$= \lim_{n\to\infty} \sum_{j=1}^{p} \mu(T^{-n}(I_j) \cap J_k)$$
$$= \sum_{j=1}^{p} \lim_{n\to\infty} \mu(T^{-n}(I_j) \cap J_k)$$
$$= \sum_{j=1}^{p} \mu(I_j)\mu(J_k) = \mu(\bigsqcup_{j=1}^{p} I_j)\mu(J_k)$$
$$= \mu(E)\mu(J_k).$$

6.3. Approximation

So,

$$\lim_{n\to\infty} \mu(T^{-n}(E) \cap F) = \lim_{n\to\infty} \mu(T^{-n}(E) \cap (\bigsqcup_{k=1}^{q} J_k))$$
$$= \lim_{n\to\infty} \sum_{k=1}^{q} \mu(T^{-n}(E) \cap J_k)$$
$$= \sum_{k=1}^{q} \lim_{n\to\infty} \mu(T^{-n}(E) \cap J_k)$$
$$= \sum_{k=1}^{q} \mu(E)\mu(J_k)$$
$$= \mu(E)\mu(\bigsqcup_{k=1}^{q} J_k) = \mu(E)\mu(F).$$

Therefore, there exists $N > 0$ so that

$$|\mu(T^{-n}(E) \cap F) - \mu(E)\mu(F)| < \varepsilon$$

for all $n > N$. Then Lemma 6.3.2 completes the proof. □

We give an application to a characterization of the weak mixing property. We saw in Proposition 6.2.2 that if T is weakly mixing, then for each pair of measurable sets A, B there exists a density zero set $D = D(A, B)$ so that $\lim_{n\to\infty, n\notin D} \mu(T^{-n}(A) \cap B) = \mu(A)\mu(B)$. Suppose that the probability space $(X, \mathcal{S}, \mu)$ has a countable sufficient semi-ring $\mathcal{R}$ (measure spaces with this property are called *separable*). So for each pair A_i, A_j in $\mathcal{R}$ there exists a density zero set $D_{i,j} = D(A_i, A_j)$ outside of which the transformation is mixing (i.e., $\lim_{n\to\infty, n\notin D_{i,j}} \mu(T^{-n}(A_i) \cap A_j) = \mu(A_i)\mu(A_j)$). This means that for each pair of positive integers i, j there exists an infinite sequence $n_k(i, j)$ such that

$$\lim_{k\to\infty} \mu(T^{-n_k(i,j)}(A_i) \cap A_j) = \mu(A_i)\mu(A_j).$$

As the semi-ring $\mathcal{R}$ is countable, a diagonalization argument yields a sequence $\{n_k\}$ such that for all i, j,

$$\lim_{k\to\infty} \mu(T^{-n_k}(A_i) \cap A_j) = \mu(A_i)\mu(A_j)$$

(Exercise 2). The following lemma shows that this sequence works for all elements of $\mathcal{S}$.

Lemma 6.3.5. *Let $(X, \mathcal{S}, \mu)$ be a probability space. Suppose that there exists a sufficient semi-ring $\mathcal{R}$ and a sequence $\{n_k\}$ so that for all elements I and J in $\mathcal{R}$,*
$$\lim_{k \to \infty} \mu(T^{-n_k}(I) \cap J) = \mu(I)\mu(J).$$
Then for all A and B in $\mathcal{S}$,
(6.4) $$\lim_{k \to \infty} \mu(T^{-n_k}(A) \cap B) = \mu(A)\mu(B).$$

A sequence $\{n_k\}$ satisfying (6.4) is said to be a **mixing sequence** for the transformation T.

Proof. The proof that $\{n_k\}$ is a mixing sequence for all measurable sets is as in Theorem 6.3.4. $\square$

While from our construction it does not necessarily follow that the sequence $\{n_k\}$ is of density one, it can be shown that the mixing sequence $\{n_k\}$ may be chosen to be of density one.

Exercises

(1) Prove Lemma 6.3.3.

(2) Let $(X, \mathcal{S}, \mu)$ be a probability space with a countable sufficient semi-ring $\mathcal{R}$. Show that if T is weakly mixing, then there exists a sequence $\{n_k\}$ that is mixing for all elements I and J in $\mathcal{R}$.

* (3) Let $(X, \mathcal{S}, \mu)$ be a probability space with a countable sufficient semi-ring. Show that if T is weakly mixing, then there exists a sequence $\{n_k\}$ of density one that is mixing.

6.4. Characterizations of Weak Mixing

This section studies the weak mixing property in more detail. We start with a characterization of weak mixing in terms of Cartesian products.

Proposition 6.4.1. *Let T be a measure-preserving transformation on a probability space $(X, \mathcal{S}, \lambda)$. Then the following are equivalent.*

6.4. Characterizations of Weak Mixing

(1) T is weakly mixing.

(2) $T \times T$ is weakly mixing.

(3) $T \times T$ is ergodic.

Proof. Suppose that T is weakly mixing. Let A, B, C, D be measurable sets in X. Then there exist zero density sets D_1 (for A and B) and D_2 (for C and D) so that for all $n \notin (D_1 \cup D_2)$,

$$\lim_{n \to \infty} \lambda(T^{-n}(A) \cap B) = \lambda(A)\lambda(B),$$
$$\lim_{n \to \infty} \lambda(T^{-n}(C) \cap D) = \lambda(C)\lambda(D).$$

Let λ^2 denote the product measure $\lambda \times \lambda$ on $X \times X$. Then, for all $n \notin (D_1 \cup D_2)$,

(6.5)
$$\lim_{n \to \infty} \lambda^2[(T \times T)^{-n}(A \times C) \cap (B \times D)] = \lambda^2(A \times C)\lambda^2(B \times D).$$

Since (6.5) holds for all elements of a sufficient semi-ring, Lemma 6.3.5 implies that (6.5) holds for all measurable sets. Then by Proposition 6.2.2, $T \times T$ is weakly mixing, and also $T \times T$ is ergodic.

Now suppose that $T \times T$ is ergodic. We will prove that T satisfies Proposition 6.2.2(3). So we evaluate, for any A, B that are measurable sets in X,

$$\lim_{n \to \infty} \frac{1}{n} \sum_{i=0}^{n-1} \left(\lambda(T^{-i}(A) \cap B) - \lambda(A)\lambda(B) \right)^2$$
$$= \lim_{n \to \infty} \frac{1}{n} \sum_{i=0}^{n-1} (\lambda(T^{-i}(A) \cap B))^2$$
$$- 2 \lim_{n \to \infty} \frac{1}{n} \sum_{i=0}^{n-1} (\lambda(T^{-i}(A) \cap B))\lambda(A)\lambda(B) + (\lambda(A)\lambda(B))^2.$$

Since $T \times T$ is ergodic, then T is ergodic (see Exercise 6.4.1), so by Theorem 5.1.5,

$$\lim_{n \to \infty} \frac{1}{n} \sum_{i=0}^{n-1} \lambda(T^{-i}(A) \cap B) = \lambda(A)\lambda(B).$$

As $T \times T$ is ergodic, applying Theorem 5.1.5 to the sets $A \times A$ and $B \times B$,

$$\lim_{n \to \infty} \frac{1}{n} \sum_{i=0}^{n-1} \lambda^2[(T \times T)^{-i}(A \times A) \cap (B \times B)] = \lambda^2(A \times A)\lambda^2(B \times B),$$

so,

$$\lim_{n \to \infty} \frac{1}{n} \sum_{i=0}^{n-1} \left(\lambda(T^{-i}(A)) \cap (B)\right)^2 = \lambda(A)^2 \lambda(B)^2.$$

Therefore,

$$\lim_{n \to \infty} \frac{1}{n} \sum_{i=0}^{n-1} \left(\lambda(T^{-i}(A) \cap B) - \lambda(A)\lambda(B)\right)^2 = 0.$$

Proposition 6.2.2 shows that T is weakly mixing. □

We conclude with some additional properties that are equivalent to weak mixing. The hard part of the proof of Theorem 6.4.2 is to show that if T has continuos spectrum, then $T \times S$ is ergodic for any ergodic finite measure-preserving S (i.e., (2) $\Rightarrow$ (4) below). The proof of this fact typically uses the spectral theorem, which is beyond the scope of this book (for a proof see, e.g., [57]). We give the outline of a different proof that also uses concepts not covered in this book; the reader should use our outline as a guide to further explore these interesting topics. We give complete proofs for the directions that use notions developed in this book.

Theorem 6.4.2. *Let T be an invertible measure-preserving transformation on a Lebesgue probability space. Then the following are equivalent.*

(1) *T is weakly mixing.*
(2) *T has continuous spectrum.*
(3) *T is doubly ergodic.*
(4) *$T \times S$ is ergodic for any ergodic, finite measure-preserving transformation S.*

Proof. (1) $\Rightarrow$ (2): Suppose first that T is weakly mixing. Let $f \in L^2$ be an eigenfunction with eigenvalue λ and $|f| = 1$. Define $g : X \times X \to$

6.4. Characterizations of Weak Mixing 217

$\mathbb{C}$ by $g(x,y) = f(x)\bar{f}(y)$. Then
$$g(T(x), T(y)) = f(T(x))\bar{f}(T(y)) = \lambda f(x)\bar{\lambda}\bar{f}(y)$$
$$= f(x)\bar{f}(y) = g(x,y).$$
As $T \times T$ is ergodic, then g is constant a.e., a contradiction unless f is constant a.e.

(1) $\Rightarrow$ (3): This is done in Proposition 6.2.3.

(2) $\Rightarrow$ (4): Now we outline a proof that if T has continuous spectrum, then $T \times S$ is ergodic for any ergodic finite measure-preserving invertible transformation S. We follow [**63**, Theorem 4.10]; for another approach see [**58**, Section 4.1]. If $T \times S$ were not ergodic, then there would exist a set $A \subset X \times Y$ of positive measure such that $T \times S(A) = A$. Define, for $x \in X$, the fiber sets A_x to be
$$A_x = \{y \in Y : (x,y) \in A\}.$$
The fact that A is $T \times S$-invariant implies that $S(A_x) = A_{T(x)}$ for a.e. x, so S takes the fiber at x to the fiber at $T(x)$. It can also be shown that $\nu(A_x) = \mu \times \nu(A)$ for a.e. x. We define a metric by
$$d(x, x') = \nu(A_x \triangle A_{x'}),$$
for all x, x' in a T-invariant set of full measure in X. One verifies that d is a metric after we identify points by $x \sim x'$ if and only if $A_x = A_{x'}$ a.e. Then $X/\sim$ with d is a metric space. We claim that T acts on $X/\sim$ as an isometry and that $X/\sim$ with this metric is compact. Compactness follows from verifying that the metric is totally bounded and complete. It follows that T on $X/\sim$ is isomorphic to a rotation on a compact group. It can be shown that a rotation on a compact group has a nonconstant eigenfunction, contradicting the fact that T has continuous spectrum.

(3) $\Rightarrow$ (2): T is clearly ergodic. Let λ be an eigenvalue for T. So there exists a function $f \in L(X, \mu, \mathbb{C})$ such that $f(T(x)) = \lambda f(x)$ a.e. and $|\lambda| = 1, |f| = 1$. We could work on the unit circle but for convenience we shall pass to the unit interval. Write $\lambda = e^{2\pi i \alpha}$ and $f(x) = e^{2\pi i g(x)}$ for some $\alpha \in [0,1)$ and some measurable function $g : X \to [0,1)$. Let $R : [0,1) \to [0,1)$ be defined by $R(t) = t + \alpha$. Then $g \circ T = R \circ g$. Define a probability measure ν on $[0,1)$ by $\nu(A) = \mu(g^{-1}(A))$. Then g is a factor map from T to R. As R is a

factor it must be ergodic. If α is rational, then the measure ν must be atomic and concentrated on a finite number of atoms; in this case R is not doubly ergodic, a contradiction. If α is irrational, then ν must be Lebesgue measure. In this case also R cannot be doubly ergodic, a contradiction. Therefore $\lambda = 1$.

(4) $\Rightarrow$ (1): This follows from Proposition 6.4.1.

$\square$

Exercises

(1) Show that if $T \times T$ is ergodic, then T^k, $k > 0$, is ergodic (without using Proposition 6.4.1), and consider the case when T could possibly be infinite measure-preserving.

(2) For any sets A, B, C, D show that $(A \times B) \cap (C \times D) = (A \cap C) \times (B \cap D)$.

(3) Show that if T is weakly mixing, then T^n is weakly mixing for $n > 1$.

6.5. Chacón's Transformation

This is an example of a weakly mixing transformation that is not mixing. The construction of Chacón's transformation is similar to the construction of the dyadic odometer and the Hajian-Kakutani transformation. The transformation is defined by a sequence of columns. As in the case of the Hajian-Kakutani transformation, the columns have increasing measure, but in this case the measure converges to a finite number, so that the resulting transformation is finite measure-preserving. Start by letting C_0 consist of the single level $I_{0,0} = [0, \frac{2}{3})$ and of height $h_0 = 1$. For concreteness we show how column C_1 is constructed. First cut $I_{0,0}$ into the subintervals $[0, \frac{2}{9}), [\frac{2}{9}, \frac{4}{9}), [\frac{4}{9}, \frac{2}{3})$. Place a spacer (a new level) above the middle subinterval. This new level is chosen to abut the current interval and is the interval $[\frac{2}{3}, \frac{8}{9})$. (The construction is such that at any stage the union of the levels of a column equals an interval.) Place the interval $[\frac{2}{3}, \frac{8}{9})$ above $[\frac{2}{9}, \frac{4}{9})$ and stack from left to right to obtain the column on the right of Figure 6.1.

For the inductive step, assume that column C_n of height h_n has been constructed. To obtain column C_{n+1}, cut each level of C_n into

6.5. Chacón's Transformation

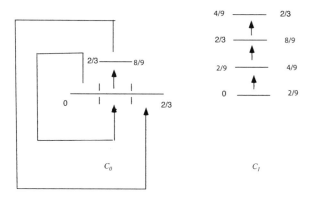

Figure 6.1. Generating C_1 from C_0 for Chacón's transformation

three equal subintervals to produce three subcolumns of C_n, numbered from left to right as $C_{n,i}, i = 0, 1, 2$. Then place a spacer (a new level) above the top of $C_{n,1}$, of the same length as the top level of $C_{n,1}$, and stack the three subcolumns from left to right, i.e., the top level of $C_{n,0}$ is sent by a translation map to the bottom level of $C_{n,1}$, the top level of $C_{n,1}$ is sent to the spacer, and the spacer at the top of $C_{n,1}$ is sent by a translation map to the bottom of $C_{n,2}$. This defines a new column C_{n+1} of height $h_{n+1} = 3h_n + 1$; levels in C_{n+1} are numbered from bottom to top. On each level the transformation is defined by the (unique orientation-preserving) translation map that maps that level onto the one on top. The total measure of the spacers is

$$2/9 + 2/27 + \ldots = \frac{2/9}{1 - 1/3} = \frac{1}{3},$$

so this defines a transformation on $[0, 1)$, and we call this transformation the **canonical Chacón transformation**, or Chacón's transformation for short. See Figure 6.2. Chacón's transformation was constructed in 1969 and is one of the early examples of a weakly mixing transformation that is not mixing; it is also the source of several interesting examples and counterexamples. For example, del Junco showed in 1978 that Chacón's transformation T only commutes with its powers, i.e., if S is a measure-preserving transformation such that $S \circ T = T \circ S$ a.e., then $S = T^k$ for some integer k.

Suppose $I = T^j(I_{\ell,0})$ is in C_ℓ, for $j = 0, \ldots, h_\ell - 1$. For any $n > \ell$, I is the union of some elements in

$$C_n = \{I_{n,0}, T(I_{n,0}), \ldots, T^{h_n-1}(I_{n,0})\}.$$

We refer to the elements in this union as **copies** or **sublevels** of I. For example, any level I in C_ℓ has three copies in $C_{\ell+1}$, each of length $\frac{1}{3}|I|$, and it has 3^2 copies in $C_{\ell+2}$, each of length $\frac{1}{9}|I|$.

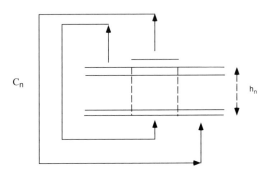

Figure 6.2. C_n-column for Chacón's transformation

We start with a lemma that shows the combinatorial dynamics of levels in Chacón's transformation. It will be used to show several properties of the transformation. We say levels I and J in C_n are $|\ell|$ apart if J is the $i+\ell$ level in C_n when I is the i level of C_n. (There is a stronger version of the lemma that is left as an exercise (Exercise 3), but is not needed in our proofs.)

Lemma 6.5.1. *Let $n > 0$ and let I and J be levels in C_n.*

(1) *For all $k \geq n$,*

$$\lambda(T^{h_k}(I) \cap I) \geq \frac{1}{3}\lambda(I).$$

(2) *If I and J are at most $\ell \geq 0$ apart, with I above J, then there exists an integer $H = H(n, \ell)$ such that*

$$\lambda(T^H(I) \cap J) \geq (\frac{1}{3})^\ell \lambda(J).$$

6.5. Chacón's Transformation

Proof. First let $k = n$. For any level K in C_n let the three copies of K in C_{n+1} be denoted by $K[0], K[1], K[2]$, respectively. Then from the construction of T one observes that

(6.6) $\quad T^{h_n}(K[0]) = K[1], \quad T^{h_n}(K[1]) = (T^{-1}(K))[2].$

So $T^{h_n}(I)$ intersects both I and $T^{-1}(I)$ in measure at least $\frac{1}{3}$ times the measures of I and $T^{-1}(I)$, respectively (we discard smaller pieces):

(6.7) $\quad\quad\quad\quad\quad\quad \lambda(T^{h_n}(I) \cap I) \geq \frac{1}{3}\lambda(I),$

(6.8) $\quad\quad\quad\quad\quad\quad \lambda(T^{h_n}(I) \cap T^{-1}(I)) \geq \frac{1}{3}\lambda(I).$

For $k = n + \ell, \ell > 0$, each I in C_n has 3^ℓ copies in C_k and, by the same reasoning as above, for each copy I' of I in C_k,

$$\lambda(T^{h_k}(I') \cap I') \geq (1/3)\lambda(I').$$

Putting all of these copies together completes the proof of (1).

For part (2), we use that $T^{h_n}(I)$ intersects $T^{-1}(I)$ in measure at least $\frac{1}{3}$ times $\lambda(T^{-1}(I)) = \lambda(I)$, in addition to intersecting I. Since $T^{h_n}(I)$ contains two (full) levels in C_{n+1}, applying (6.6) to each of these levels gives that $T^{h_n + h_{(n+1)}}(I)$ intersects I and $T^{-1}(I)$ and $T^{-2}(I)$ in measure at least $(\frac{1}{3})^2$ times the measure of each of the respective intervals. Induction shows that

$$H = \sum_{i=0}^{\ell-1} h_{n+i}$$

satisfies the lemma. $\square$

Theorem 6.5.2. *The canonical Chacón transformation is a measure-preserving transformation on a probability Lebesgue space that has continuous spectrum.*

Proof. The proof that T is ergodic and measure-preserving is the same as that for the dyadic odometer. We show that T has continuous spectrum. Let $f \in L^2$ be an eigenfunction with eigenvalue λ and $|f| = 1$. Let $\varepsilon > 0$. There is a constant c so that the set

$$A = \{x : |f(x) - c| < \varepsilon\}$$

has positive measure (see Exercise 4.8.1). So there exists a level I in some column C_n so that $\lambda(I \cap A) > \frac{5}{6}\lambda(I)$ (by Lemma 3.7.3). Note that $T^{h_n}(I[0]) = I[1]$ and $T^{h_n+1}(I[1]) = I[2]$. Then there is a point $x \in A \cap I$ such that $T^{h_n}(x) \in A \cap I$ and $T^{h_n+1}(T^{h_n}(x)) \in A \cap I$. So

$$|f(x) - c| < \varepsilon, \quad |\lambda^{h_n} f(x) - c| < \varepsilon, \quad |\lambda^{2h_n+1} f(x) - c| < \varepsilon.$$

Therefore $|\lambda^{h_n} - 1| < 2\varepsilon$ and $|\lambda^{2h_n+1} - 1| < 2\varepsilon$. This implies that $\lambda = 1$, and therefore T has continuous spectrum. □

Lemma 6.5.3. *Chacón's transformation is not mixing.*

Proof. Let $n > 0$ be such that if I is a level in C_n, then $\lambda(I) < \frac{1}{3}$. Then for all $k \geq n$,

$$\lambda(T^{h_k}(I) \cap I) \geq \frac{1}{3}\lambda(I) > \lambda(I)\lambda(I).$$

So it cannot be the case that $\lim_{n \to \infty} \lambda(T^n(I) \cap I) = \lambda(I)\lambda(I)$. Therefore T is not mixing. □

We now present a direct proof that Chacón's transformation T is doubly ergodic. The proof that we give, with only technical modifications, can be generalized to give a direct proof that $T \times T$ is ergodic.

In the proof that Chacón's transformation is doubly ergodic we will use the following approximation lemma. The lemma can be thought of as a kind of "pigeon-hole principle" in the measurable category. We call it the Double Approximation Lemma as it is used in making a second approximation. We need some notation. A **triadic interval of order** n (in $[0, 1)$) is an interval of the form $[\frac{i}{3^n}, \frac{i+1}{3^n})$ for some $i \in \{0, \ldots, 3^n - 1\}$ and $n \in \mathbb{N}$. The lemma has natural generalizations that apply to other transformations.

Lemma 6.5.4 (Double Approximation). *Let A be a set of positive measure in $[0, 1)$ and let I be a triadic interval of order n that is 3/4-full of A. Then, for any $1 > \delta > 0$ there exists an integer $N > 0$ so that for every partition of I into triadic subintervals of order $n > N$, more than $1/2$ of the order n subintervals are each $(1 - \delta)$-full of A.*

6.5. Chacón's Transformation

Proof. We are given that

$$\lambda(I \setminus A) = \lambda(I) - \lambda(A \cap I) < \lambda(I) - \frac{3}{4}\lambda(I) = \frac{1}{4}\lambda(I).$$

Since we are only concerned with the part of A that is in I, for convenience let us agree to denote $A \cap I$ by A. We shall consider partitions of I into triadic subintervals of the same order $k > n$. Let us call such a partition a k-*partition*. Then I is the union of its k-partition elements for all $k > n$. We index the subintervals in a k-partition by elements of the set $P_{k-n} = \{0, \ldots, 3^{k-n} - 1\}$. (For example, if $k = n+1$, then there are 3 subintervals in a k-partition of I and they are indexed by elements of $P_1 = \{0, 1, 2\}$.)

Let α be a small positive number to be determined later, but depending only on δ. By Lemma 2.7.3, for any such α there is an integer N, which can be chosen to satisfy $N > n$, so that for any $k > N$ there is a set $Q \subset P_{k-n}$ such that $I' = \bigcup_{j \in Q} I_j$ satisfies

$$\lambda(A \triangle I') < \alpha\lambda(I).$$

Then

$$\lambda(I' \triangle I) \leq \lambda(I' \triangle A) + \lambda(A \triangle I)$$
$$< \alpha\lambda(I) + \frac{1}{4}\lambda(I) = (\alpha + \frac{1}{4})\lambda(I).$$

The subintervals of the k-partition that are $(1-\delta)$-full of A are chosen by letting

$$Q' = \{j \in Q : \lambda(I_j \setminus A) < \delta\lambda(I_j)\}.$$

It suffices to show that Q' is more than $1/2$ the size of P_{k-n}. For this, we first calculate, with $I'' = \bigcup_{j \in Q'} I_j$,

$$\delta\lambda(I' \triangle I'') = \delta\lambda(I' \setminus I'') = \sum_{j \in Q \setminus Q'} \delta\lambda(I_j)$$
$$\leq \sum_{j \in Q \setminus Q'} \lambda(I_j \setminus A) \leq \lambda(I' \triangle A).$$

Then
$$\lambda(I'' \triangle I) \le \lambda(I'' \triangle I') + \lambda(I' \triangle I)$$
$$\le \frac{1}{\delta}\lambda(I' \triangle A) + (\alpha + \frac{1}{4})\lambda(I)$$
$$< (\frac{\alpha}{\delta} + \alpha + \frac{1}{4})\lambda(I).$$

Now choose α so that $\frac{\alpha}{\delta} + \alpha + \frac{1}{4} < \frac{1}{2}$. Then $\lambda(I'' \triangle I) < \frac{1}{2}\lambda(I)$. This means that more than $\frac{1}{2}$ of the k-partition subintervals of I are in I''. □

Theorem 6.5.5. *The canonical Chacón transformation is doubly ergodic.*

Proof. Let A_1 and B_1 be two sets of positive measure. By similar arguments as before we may choose levels I_1 and J_1 in some column C_n that are $\frac{2}{3}$-full of A_1 and B_1 respectively, and such that I_1 is above (or at) J_1. Let $0 \le \ell \le h_n$ be such that they are ℓ apart. Let
$$\delta = (\frac{1}{3})^\ell.$$

By Lemma 6.5.4 applied twice, as levels of columns are triadic subintervals of the same order and $A \subset I$, there exists a column $C_k, k > n$, such that more than $1/2$ of the sublevels of I that are levels in C_k are $(1 - \frac{\delta}{3})$-full of A_1 and more than $1/2$ of the sublevels of J that are levels in C_k are $(1 - \frac{\delta}{3})$-full of B_1. It follows that there must be a $(1 - \frac{\delta}{3})$-full sublevel of I_1, call it I, and a $(1 - \frac{\delta}{3})$-full sublevel of J_1, call it J, such that both I and J are in the same C_n-copy in C_k (i.e., they must have the same index in the respective partition of I_1 and J_1). Thus I and J are at most ℓ apart in C_n, so by Lemma 6.5.1 there is an integer $H = H(k, \ell)$ such that

$$\lambda(T^H(I) \cap J) \ge (\frac{1}{3})^\ell \lambda(J) \text{ and } \lambda(T^H(I) \cap I) \ge (\frac{1}{3})^\ell \lambda(I).$$

Write $A = A_1 \cap I$ and $B = B_1 \cap J$. We also know that
$$\lambda(I \setminus A) < \frac{\delta}{3}\lambda(I) \quad \text{and}$$
$$\lambda(J \setminus B) < \frac{\delta}{3}\lambda(J) = \frac{\delta}{3}\lambda(I).$$

6.5. Chacón's Transformation

Then by Lemma 3.7.4,

$$\lambda(T^H(A) \cap B) \geq \lambda(T^H(I) \cap J) - \lambda(I \setminus A) - \lambda(J \setminus B)$$
$$\geq \delta\lambda(J) - \frac{\delta}{3}\lambda(I) - \frac{\delta}{3}\lambda(J)$$
$$> \frac{\delta}{3}\lambda(I) > 0.$$

Similarly, one obtains that for the same integer H,

$$\lambda(T^H(A) \cap A) \geq \lambda(T^H(I) \cap I) - \lambda(I \setminus A) - \lambda(I \setminus A)$$
$$> \frac{\delta}{3}\lambda(I) > 0.$$

Therefore T is doubly ergodic. $\square$

Exercises

(1) Show that the proof in Theorem 6.5.2 holds for the case of a generalized Chacón transformation where all one knows is that spacers are added infinitely often (the times when a "spacer is not added" are those n such that column C_{n+1} is obtained by cutting each level of C_n into three pieces and stacking from left to right). What if the spacer is added only a finite number of times?

(2) Let T be Chacón's transformation. Show that for any $n > 0$ and any level I of C_n,

$$T^{h_k}(I) \subset I \cup T^{-1}(I) \text{ for all } k \geq n.$$

(3) Let T be Chacón's transformation. Show that for any $n > 0$ and any level I of C_n,

$$\lambda(T^{h_k}(I) \cap I) = \frac{1}{2}\lambda(I) \text{ for all } k \geq n.$$

(4) A finite measure-preserving transformation T on a probability space is said to be **lightly mixing** if for all sets A, B of positive measure,

$$\liminf_{n \to \infty} \lambda(T^{-n}(A) \cap B) > 0.$$

Show that a lightly mixing transformation is weakly mixing.

(5) Show that (the canonical) Chacón's transformation is not lightly mixing (so weak mixing does not imply light mixing).

(6) Let T be a recurrent, ergodic measure-preserving transformation. Show that for all pairs of sets of positive measure A and B there exists an integer $n > 0$ such that $\mu(T^{-n}(A) \cap A) > 0$ and $\mu(T^{-n}(B) \cap B) > 0$.

(7) Let T be Chacón's transformation. Modify the proof of double ergodicity to show that for any sets A, B, C, D of positive measure there exists an integer $n > 0$ such that $\lambda(T^n(A) \cap B) > 0$ and $\lambda(T^n(C) \cap D) > 0$.

* (8) Let T be Chacón's transformation. Show directly, as in the proof of double ergodicity, that the Cartesian product transformation $T \times T$ on $X \times X$ is ergodic.

* (9) Prove the following generalization of the Double Approximation Lemma: Let A be a set of positive measure in $[0, 1)$ and $0 < \tau < 1$. Then, for any $1 > \delta > 0$, there exists an integer $N > 0$ and a q-adic subinterval I so that for every partition of I into q-adic subintervals of order $n > N$, more than (100τ)-percent of the order n subintervals of I are each $(1 - \delta)$-full of A.

6.6. Mixing

In this section we give an example of a mixing transformation.

Theorem 6.6.1. *Let $X = [0, 1]$ and $T(x) = 2x \bmod 1$. Then T is mixing on $(X, \mathcal{S}, \lambda)$.*

Proof. By Theorem 6.3.4, it suffices to verify the mixing condition on the sufficient semi-ring of dyadic intervals. Recall that a dyadic interval of order k has the form $[\frac{i}{2^k}, \frac{i+1}{2^k})$, for some $i = 0, \ldots, 2^k - 1$. As any two dyadic intervals of the same order are disjoint and any dyadic interval is a finite union of dyadic intervals of larger order, it suffices to show mixing for the case when A and B are dyadic intervals of the same order. One can verify by induction that if A and B are dyadic intervals of order k, $\lambda(T^{-n}(A) \cap B) = \lambda(A)\lambda(B)$ for $n > k$. Therefore T is mixing. □

Exercises

(1) Complete the proof of Theorem 6.6.1.

(2) Let $T = 10x \mod 1$ on $[0, 1)$. Show that T is mixing.

(3) Show that if T is mixing, then T^n is mixing for all $n > 0$.

(4) Show that if T is mixing, then any factor of T is mixing.

(5) Show that the shift transformation σ is mixing on Σ_N^+.

(6) Show that the baker's transformation is mixing.

* (7) (Renyi) Let T be a probability-preserving transformation. Show that if for all measurable sets A,
$$\lim_{n \to \infty} \lambda(T^{-n}(A) \cap A) = \lambda(A)^2,$$
then T is mixing.

6.7. Rigidity and Mild Mixing

We start with an interesting dynamical property that is enjoyed by rotations. A finite measure-preserving transformation T is said to be **rigid** if for all measurable sets A and for any $\varepsilon > 0$, there exists an integer $n > 0$ such that $\lambda(T^{-n}(A) \triangle A) < \varepsilon$. Clearly, the identity transformation is a rigid transformation, but we also have the following theorem.

Theorem 6.7.1. *Rotations are rigid transformations.*

Proof. Let $R = R_\alpha$ be a rotation on $[0, 1)$. If α is a rational number, then R is periodic of some period p and so $R^p(A) = A$ for all sets A and thus R is trivially rigid. Let α be irrational. First consider the case when A is an interval, say $A = (a, b)$ (our argument does not need the interval to be open). Let $\varepsilon_1 > 0$. As $\{R^k(b)\}_{k>0}$ is a dense sequence in $[0, 1)$ there exists an integer $n > 0$ such that $b < R^n(b) < b + \varepsilon_1/2$. But since R is a rotation it follows that this inequality holds if we replace b by any other point. So, in particular, we also have that $a < R^n(a) < a + \varepsilon_1/2$. This means that $\lambda(R^n(a, b) \triangle (a, b)) < \varepsilon_1$.

Now let $\varepsilon > 0$ and let A be an arbitrary set of positive measure. We know that there exists a set G^* such that G^* is a finite union of disjoint intervals and $\lambda(G^* \triangle A) < \varepsilon/3$. Write $G^* = \bigsqcup_{j=1}^{\ell} I_j$, for some intervals I_j and some integer $\ell > 0$. By the previous part, with $\varepsilon_1 = \varepsilon/3\ell$, we know that there exists an integer $k > 0$ such that $\lambda(R^k(I_1) \triangle I_1) < \varepsilon/3\ell$. Again using that R is a rotation it follows that for $j = 2, \ldots, \ell$, $\lambda(R^k(I_j) \triangle I_j) < \varepsilon/3\ell$. Using the triangle inequality (see Exercise 2.5.6), we have that

$$\begin{aligned}\lambda(R^k(A) \triangle A) &\leq \lambda(R^k(A) \triangle R^k(G^*)) \\ &\quad + \lambda(R^k(G^*) \triangle (G^*)) + \lambda(G^* \triangle A) \\ &= \lambda(A \triangle G^*) + \lambda\left(\bigsqcup_{j=1}^{\ell} R^k(I_j) \triangle \bigcup_{j=1}^{\ell} I_j\right) \\ &\quad + \lambda(G^* \triangle A) \\ &< \varepsilon/3 + \sum_{j=1}^{\ell} \lambda(R^k(I_j) \triangle I_j) + \varepsilon/3 \\ &< \frac{\varepsilon}{3} + \ell \cdot \frac{\varepsilon}{3\ell} + \frac{\varepsilon}{3} = \varepsilon.\end{aligned}$$

□

The following is left as an exercise.

Lemma 6.7.2. *A finite measure-preserving transformation T is rigid if and only if there is a sequence $n_k \to \infty$ such that*

$$\lim_{k \to \infty} \lambda(T^{-n_k}(A) \triangle A) = 0,$$

for all measurable sets A.

There are many transformations that are not rigid but still exhibit similar behavior. This is captured by the following definition. We say that a measure-preserving transformation (on a finite or infinite measure space) S is **partially rigid** if there exists a constant $\alpha > 0$ and an increasing sequence r_n such that for all sets A of finite measure

(6.9) $$\liminf_{n \to \infty} \mu(S^{-r_n} A \cap A) \geq \alpha \mu(A).$$

Lemma 6.7.3. *Let S be a finite measure-preserving transformation satisfying (6.9) for all sets A in a dense algebra. Then S is partially rigid along the same sequence r_n.*

6.7. Rigidity and Mild Mixing

Proof. Let A be a set of finite measure. There exists a set E which is in the dense algebra and such that $\mu(A \triangle E) < \frac{\alpha}{3}$. After applying the triangle inequality, we obtain

$$\mu([S^{-r_n}(A) \cap A] \triangle [S^{-r_n}E \cap E])$$
$$\leq \mu(S^{-r_n}A \cap (A \triangle E)) + \mu(S^{-r_n}(A \triangle E) \cap E)$$
$$\leq \mu(A \triangle E) + \mu(S^{-r_n}(A \triangle E))$$
$$\leq \mu(A \triangle E) + \mu(A \triangle E) = 2\mu(A \triangle E).$$

Finally we observe that

$$\mu(S^{-r_n}A \cap A) \geq \alpha\mu(E) - 2\mu(A \triangle E).$$

We obtain the desired result by taking the limit infimum on the left-hand side and choosing a sequence of sets E arbitrarily close to A. □

Lemma 6.7.4. *Let T be a transformation on a nonatomic probability space $(X, \mathcal{S}, \mu)$. If T is partially rigid, then T is not mixing.*

Proof. Let $\{n_i\}$ be the sequence along which T is partially rigid with rigidity constant $\alpha > 0$. Let A be a measurable set such that $0 < \mu(A) < \alpha/2$. If T were mixing, then we would have $\lim_{n \to \infty} \mu(T^{-n}(A) \cap A) = \mu(A)^2$. But for infinitely many $i > 0$, $\mu(T^{-n}(A) \cap A) \geq \alpha\mu(A) > 2\mu(A)^2$, a contradiction. □

We say that a measure-preserving transformation T on a probability space is **mildly mixing** if

$$\liminf_{n \to \infty} \mu(A \triangle T^{-n}(A)) > 0$$

for all sets A with $0 < \mu(A) < 1$.

Proposition 6.7.5. *Let T be a measure-preserving transformation on a probability space. T is mildly mixing if and only if*

(6.10) $$\liminf_{n \to \infty} \mu(A^c \cap T^{-n}(A)) > 0$$

for all sets A with $0 < \mu(A) < 1$.

Proof. Using the following equality from set theory

$$A \triangle T^{-n}(A) = [A^c \cap T^{-n}A] \cup [A \cap T^{-n}A^c],$$

if (6.10) holds, then $\liminf_{n\to\infty} \mu(A \triangle T^{-n}(A)) > 0$, so T is mildly mixing.

Now let T be mildly mixing and suppose there is a sequence $\{n_i\}$ increasing to ∞, such that

(6.11) $$\lim_{i\to\infty} \mu(A^c \cap T^{-n_i} A) = 0.$$

Since T is measure-preserving,

$$\mu(A) = \mu(T^{-n_i}(A)) = \mu(A^c \cap T^{-n_i}(A)) + \mu(A \cap T^{-n_i} A).$$

Then

$$\mu(A) = \lim_{i\to\infty} \mu(A \cap T^{-n_i} A).$$

Now, using that $\mu(A) = \mu(A \cap T^{-n_i} A) + \mu(A \cap T^{-n_i} A^c)$, we obtain

$$\lim_{i\to\infty} \mu(A \cap T^{-n_i} A^c) = 0.$$

Together with (6.11) this implies

$$\lim_{i\to\infty} \mu(A \triangle T^{-n_i} A) = 0,$$

a contradiction. Therefore (6.10) holds. □

Exercises

(1) Show that a mixing transformation is mildly mixing.

(2) Show from the definition that a mildly mixing transformation is ergodic.

(3) Show that the dyadic odometer is rigid.

* (4) (Friedman) Show that Chacón's transformation is partially rigid with $\alpha = 2/3$. (A simpler exercise is to simply show that T is partially rigid.)

(5) Construct a weakly mixing transformation that is rigid.

(6) Let k be a fixed positive integer and let $r_n = k$ for all $n \geq 0$. Show that the r_n-adic odometer is a rigid, ergodic measure-preserving transformation. Show that T^k is not ergodic.

(7) Show that the r_n-adic odometer for $r_n = n$ is a rigid, ergodic measure-preserving transformation.

6.8. When Approximation Fails

This section presents examples of transformations defined on the unit interval that satisfy strong dynamical properties on all intervals but do not satisfy these properties on all measurable sets. The constructions are based on a set K^* of the unit interval that is constructed in the appendix using Cantor sets of positive measure. It satisfies the following interesting property. For a proof refer to Proposition B.1.8.

Lemma 6.8.1. *There exists a Borel set K^* of the unit interval such that for every nonempty open interval I in $[0,1]$, $\lambda(I \cap K^*) > 0$ and $\lambda(I \cap K^{*c}) > 0$.*

A finite measure-preserving transformation T is **lightly mixing** if for all sets A and B of positive measure

$$\liminf_{n \to \infty} \mu(T^{-n} A \cap B) > 0.$$

A mixing transformation must be lightly mixing, and it is clear that a lightly mixing transformation is doubly ergodic, so weakly mixing. For us here it suffices to think of light mixing as a condition stronger than weak mixing.

The transformation T is said to be **lightly mixing on an algebra** $\mathcal{A}$ if $\liminf_{n \to \infty} \mu(T^{-n} A \cap B) > 0$ for all sets of positive measure $A \in \mathcal{A}$ and $B \in \mathcal{S}$. For example, if a transformation is lightly mixing on an algebra $\mathcal{A}$, it cannot admit invariant sets of positive measure, with positive measure complement, in $\mathcal{A}$.

We first construct a finite measure-preserving transformation T that is not ergodic but is lightly mixing on the algebra of (finite unions of) intervals, and then an ergodic transformation S that is lightly mixing on the algebra of intervals but such that S^2 is not ergodic.

The following proposition is similar to the second example in Section 3.10 and its proof is left to the reader.

Proposition 6.8.2. *The functions $\phi : K^* \to [0, 1/2]$ and $\psi : K^{*c} \to [0, 1/2]$ defined by $\phi(x) = \mu(K^* \cap [0,x))$ and $\psi(x) = \mu(K^{*c} \cap [0,x))$ have well-defined inverses a.e. and are measure-preserving.*

Proposition 6.8.2 is our first example showing that it is not sufficient to verify a condition such as ergodicity on a dense algebra.

Proposition 6.8.3. *There exists a finite measure-preserving transformation T that is lightly mixing on the algebra of intervals but is not ergodic.*

Proof. Let T_0 be a lightly mixing invertible transformation on $[0,1)$ (for example, any mixing transformation). First define the two transformations $T_1 : K^* \to K^*$ and $T_2 : K^{*c} \to K^{*c}$ by $T_1 = \phi^{-1} \circ T_0 \circ \phi$ and $T_2 = \psi^{-1} \circ T_0 \circ \psi$. Since ϕ, ψ and T_0 are measure-preserving and invertible, T_1 and T_2 are measure-preserving as well. In addition, both T_1 and T_2 are clearly lightly mixing on K^* and K^{*c} respectively, since they are isomorphic to T_0. Now define T by

$$T(x) = \begin{cases} T_1(x), & \text{for } x \in K^*, \\ T_2(x), & \text{for } x \in K^{*c}. \end{cases}$$

As T has an invariant set of positive measure, it is not ergodic. To show it is lightly mixing on intervals first write, for any set A, $A_1 = A \cap K^*$ and $A_2 = A \cap K^{*c}$. We know that if I is an interval of positive length in $[0,1]$, then $\mu(I \cap K^*) > 0$ and $\mu(I \cap K^{*c}) > 0$. It is clear that

$$T^{-n} I \cap A = (T_1^{-n} I_1 \cap A_1) \cup (T_2^{-n} I_2 \cap A_2).$$

As T_1 and T_2 are lightly mixing and either $\mu(A_1) > 0$ or $\mu(A_2) > 0$ it follows that

$$\liminf_{n \to \infty} \mu(T^{-n} I \cap A) > 0.$$

This shows that T is lightly mixing on intervals. □

Our next example constructs an ergodic transformation S that is lightly mixing on the a dense algebra but is not weakly mixing (S^2 is not ergodic).

Proposition 6.8.4. *There exists an ergodic finite measure-preserving transformation S that is lightly mixing on the algebra of intervals but such that S^2 is not ergodic.*

Proof. First define the transformations $S_1 : K^* \to K^{*c}$ and $S_2 : K^{*c} \to K^*$ by $S_1 = \psi^{-1} \circ T_0 \circ \phi$ and $S_2 = \phi^{-1} \circ T_0 \circ \psi$. As is

6.8. When Approximation Fails

the case with T_1 and T_2, S_1 and S_2 are measure-preserving. Now let $S : [0,1] \to [0,1]$ be defined by

$$S(x) = \begin{cases} S_1(x), & \text{for } x \in K^*, \\ S_2(x), & \text{for } x \in K^{*c}. \end{cases}$$

Evidently, K^* is an invariant set for S^2, so it is not ergodic. Now let I be an interval of positive length and let A be a set of positive measure. Suppose that $\mu(A \cap K^*) > 0$ and let $A' = A \cap K^*$. Then

$$\liminf_{n \to \infty} \mu(S^{-n} I \cap A)$$

$$\geq \min\{\liminf_{n \to \infty} \mu(S^{-(2n)} I \cap A'),$$
$$\liminf_{n \to \infty} \mu(S^{-(2n+1)} I \cap A')\}$$
$$\geq \min\{\liminf_{n \to \infty} \mu(S^{-(2n)}(I \cap K^*) \cap A'),$$
$$\liminf_{n \to \infty} \mu(S^{-(2n)}(S(I) \cap K^*) \cap A')\}.$$

As

$$S^2(x) = \begin{cases} \phi^{-1} \circ T_0^2 \circ \phi(x), & \text{for } x \in K^*, \\ \psi^{-1} \circ T_0^2 \circ \psi(x), & \text{for } x \in K^{*c}, \end{cases}$$

S^2 is lightly mixing on each of K^* and K^{*c}. Therefore

$$\liminf_{n \to \infty} \mu(S^{-n} I \cap A) > 0.$$

The case when $\mu(A \cap K^*) = 0$ follows a similar argument by letting $A' = A \cap K^{*c}$. This shows that S is lightly mixing on intervals. Ergodicity of S is left as an exercise. $\square$

Exercises

(1) Show that for any rotation by an irrational angle there is no algebra on which the transformation is lightly mixing.

(2) One can consider the weaker notion of lightly mixing between intervals: $\liminf_{n \to \infty} \mu(T^n I \cap J) > 0$ for all intervals I, J. Show that one can construct a transformation S that is lightly mixing between intervals but will not sweep out on all sequences for intervals, hence S is not lightly mixing on the algebra of intervals.

(3) Show that there exists a finite measure-preserving transformation T on $[0, 1)$ such that for all intervals I, J, there exists an integer $n > 0$ such that $\lambda(T^n(I) \cap I) > 0$ and $\lambda(T^n(I) \cap J) > 0$ but T is not ergodic.

(4) Show that there exists a finite measure-preserving transformation T on $[0, 1)$ such that for all intervals I, J, there exists an integer $n > 0$ such that $\lambda(T^n(I) \cap I) > 0$ and $\lambda(T^n(I) \cap J) > 0$ with T ergodic but not weakly mixing.

(5) (Maharam) Show that there is an ergodic measure-preserving transformation T of $[0, 1]$ such that T is not continuous when restricted to any null set.

(6) Complete the proof that S is ergodic in Proposition 6.8.4.

Open Question C. This is an open question due to Rohlin. The notion of mixing that we have defined is also called two-fold mixing. A measure-preserving transformation on a probability space is said to be *three-fold mixing* if or all measurable sets A, B and C

$$\mu(A \cap T^n(B) \cap T^m(C)) \to \mu(A)\mu(B)\mu(C)$$

as $n, m \to \infty$ and $n - m \to \infty$. In a similar way one defines k-fold mixing for $k > 3$. It is clear that three-fold mixing implies two-fold mixing. It is not known whether two-fold mixing implies k-fold mixing for all $k > 2$, or even if it implies three-fold mixing. This question dates from a paper of Rohlin's published in 1949. It is not even known whether two-fold mixing implies that

$$\lim_{n \to \infty} \mu(A \cap T^n(B) \cap T^{2n}(C)) = \mu(A)\mu(B)\mu(C),$$

for all measurable sets A, B and C. a) Show that the doubling map is k-fold mixing of all $k \geq 2$. b) Find the definition of K-automorphisms and show that K-automorphisms are mixing of all orders. c) Describe the advances on this question due to Kalikow, Ryzhikov and Host. d) Do you think that two-fold mixing implies three-fold mixing? Give some reasons. e) Describe the notions of two-fold mixing and three-fold mixing for actions of $\mathbb{Z}^2$ and Ledrappier's counterexample. f) Solve the problem.

Appendix A

Set Notation and the Completeness of $\mathbb{R}$

We assume that the reader is familiar with the basic notions from set theory such as the union and intersection of sets and properties of the real numbers such as the field operations (properties of multiplication and addition) and the order axioms (properties of inequalities); for these the reader may refer to the real analysis books cited in the bibliography. The purpose of this section is to list some notation from set theory and then to review the completeness property of the real numbers, which is at the basis of what are called compactness arguments in analysis. Our presentation is complete but terse. The material in this section is covered in a typical first course in undergraduate analysis.

Set Notation

If x is an **element** of a set S we write $x \in S$. The **empty set** is denoted by $\emptyset$. A set A is a **subset** of a set B, written $A \subset B$, if whenever $x \in A$, then $x \in B$. So $\emptyset \subset A$ and $A \subset A$ for any set A. We write $A = B$ if $A \subset B$ and $B \subset A$. We write $A \subsetneq B$ when $A \subset B$ but $A \neq B$. The union of two sets A and B is denoted by $A \cup B$ and their intersection by $A \cap B$. The **symmetric difference** of two sets is $A \triangle B = (A \setminus B) \cup (B \setminus A)$.

The notions of union and intersection can be extended to the case of an arbitrary family of sets. Let Γ be a set and let $\{A_\alpha\}$ be a collection of sets in some set X indexed by Γ (i.e., there is a function that assigns to each α in Γ a set A_α). The union and intersection of the collection $\{A_\alpha\}$ are defined, respectively, by

$$\bigcup_{\alpha \in \Gamma} A_\alpha = \{x \in X : x \in A_\alpha \text{ for some } \alpha \in \Gamma\},$$

$$\bigcap_{\alpha \in \Gamma} A_\alpha = \{x \in X : x \in A_\alpha \text{ for all } \alpha \in \Gamma\}.$$

A collection of sets A_α, $\alpha \in \Gamma$, is said to be **pairwise disjoint** or **disjoint** if $A_\alpha \cap A_\beta = \emptyset$ for all $\alpha, \beta \in \Gamma$ with $\alpha \neq \beta$. To simplify the notation we write *disjoint* for *pairwise disjoint*. We speak of a **countable union** or **countable intersection** when the set Γ is countable. When $\{A_\alpha\}_{\alpha \in \Gamma}$ is a collection of pairwise disjoint sets we may write

$$\bigsqcup_{\alpha \in \Gamma} A_\alpha$$

for the union $\bigcup_{\alpha \in \Gamma} A_\alpha$ in order to make it explicit in the notation that the union is over a pairwise disjoint family.

Real Numbers

We use the following notation:

$\mathbb{N} = $ the set of positive integers,
$\mathbb{N}_0 = \mathbb{N} \cup \{0\}$,
$\mathbb{Z} = $ the set of integers,
$\mathbb{Q} = $ the set of rational numbers,
$\mathbb{R} = $ the set of real numbers,
$\mathbb{R}^* = $ the set of extended real numbers $= \mathbb{R} \cup \{+\infty, -\infty\}$,
$\mathbb{C} = $ the set of complex numbers.

A. Completeness of the Real Numbers

The algebra of $\mathbb{R}$ is extended to $\mathbb{R}^*$ with the convention that, for $x \in \mathbb{R}$,

$$x \cdot +\infty = +\infty, x \cdot -\infty = -\infty, \text{ when } x > 0,$$
$$x \cdot +\infty = -\infty, x \cdot -\infty = +\infty, \text{ when } x < 0,$$
$$x - \infty = -\infty, x + \infty = \infty, 0 \cdot \pm\infty = 0,$$
$$+\infty \cdot \pm\infty = \pm\infty, -\infty \cdot -\infty = +\infty.$$

Intervals

Given $a, b \in \mathbb{R}$ with $a \leq b$, define an **open bounded interval**, denoted (a, b), by $(a, b) = \{x \in \mathbb{R} : a < x < b\}$ (the interval (a, b) is empty when $b = a$). The **length** of this interval is denoted by $|(a, b)|$ and defined to be $|(a, b)| = b - a$. A **closed bounded interval** has the form $[a, b] = \{x \in \mathbb{R} : a \leq x \leq b\}$. **Half-open** intervals have the form $(a, b] = \{x \in \mathbb{R} : a < x \leq b\}$ and $[a, b) = \{x \in \mathbb{R} : a \leq x \leq b\}$. (We rather arbitrarily say half-open rather than half-closed.) A **bounded interval** (sometimes called a finite interval) is any interval of the form $(a, b), (a, b], [a, b), [a, b]$. **Unbounded intervals** (also called infinite intervals) may be open or closed and have the form $(-\infty, \infty)$, $(-\infty, b), (a, \infty), (-\infty, b], [a, \infty)$ and are defined in a similar way to the bounded intervals. The length of an unbounded interval is ∞. By an **interval** we mean any unbounded or bounded interval (possibly empty). For any $a \in \mathbb{R}$ and any number $\varepsilon > 0$, the **open ball** around a of radius ε is denoted by $B(a, \varepsilon)$ and defined by $B(a, \varepsilon) = \{x \in \mathbb{R} : |x - a| < \varepsilon\}$; this is nothing but an open interval of length 2ε centered at a.

Greatest Lower Bound and Least Upper Bound

An important theme of Chapter 2 is to approximate sets using sequences of intervals. The notion of infimum will be crucial to this approximation. This is also the important notion in our formulation of the completeness property. If S is any nonempty set of real numbers, an **upper bound** for S is a number b such that for all $t \in S$, $t \leq b$; similarly a **lower bound** c satisfies $c \leq t$ for all $t \in S$. A set S is said to be **bounded below** if it has a lower bound and **bounded**

above if it has an upper bound. S is **bounded** if it is bounded above and bounded below.

The **infimum** or **greatest lower bound** of S, denoted $\inf S$, is a number α that is a lower bound for S and such that if b is any other lower bound for S, then $b \leq \alpha$. The **supremum** or **least upper bound** of S, denoted $\sup S$, is a number β that is an upper bound for S and such that if c is any other upper bound for S, then $c \geq \beta$. The following lemma gives a useful characterization of the infimum of a set and is used often; a similar statement holds for the supremum.

Lemma A.1.1. *If S is any nonempty set of numbers, a real number α is the infimum of S if and only if α is a lower bound for S and for any $\varepsilon > 0$ there is $t \in S$ such that $t < \alpha + \varepsilon$.*

Proof. Let α be the infimum of S. Clearly α is a lower bound for S. Now let $\varepsilon > 0$. Suppose that for every $t \in S$, $t \geq \alpha + \varepsilon$. Then $\alpha + \varepsilon$ is also a lower bound for S, but this contradicts that α is the greatest lower bound. Therefore there is a $t \in S$ with $t < \alpha + \varepsilon$. For the converse suppose that b is any lower bound for S. If $b > \alpha$, then let $\varepsilon = b - \alpha > 0$. So there exists $t \in S$ such that $t < \alpha + \varepsilon$, or $t < b$, which contradicts that b is a lower bound for S. Therefore $b \leq \alpha$, which together with the fact that α is a lower bound implies that $\alpha = \inf S$. $\square$

Completeness

The following fundamental property is at the heart of the characterization of the real numbers. In our presentation we assume it is an axiom of the real numbers.

Completeness Property of the Reals. *Any nonempty set of real numbers that has a lower bound has a greatest lower bound, or infimum.*

It will be convenient to extend the definition of infimum and supremum to all sets. If S is nonempty and unbounded below we write $\inf S = -\infty$ and if unbounded above, $\sup S = \infty$. Furthermore,

A. Completeness of the Real Numbers 239

we set inf $\emptyset = \infty$ and sup $\emptyset = -\infty$ (we think of every real number as a lower bound, and also an upper bound, for $\emptyset$).

Exercise. State the analogue of the completeness property in the case of the supremum. Using the fact that for any set S, sup $S = -\inf\{-S\}$ (where $-S = \{-t : t \in S\}$) show that the Completeness Property for the infimum is equivalent to the Completeness Property for the supremum.

The following fundamental theorem is now a consequence of the Completeness Property. We leave it as an exercise to complete the details of the proof. Recall that a sequence $\{x_n\}_{n \geq 1}$ is said to **converge** to $x \in \mathbb{R}$, written as $\lim_{n \to \infty} x_n = x$, if for all $\varepsilon > 0$ there is an integer $N = N(\varepsilon) \geq 1$ such that $|x - x_n| < \varepsilon$ for all $n \geq N$.

Theorem A.1.2 (Bolzano–Weierstrass). *Any bounded sequence of real numbers has a convergent subsequence.*

Proof. Let $\{x_n\}$ be a bounded sequence in $\mathbb{R}$. Any sequence $\{x_n\}$ in $\mathbb{R}$ contains a subsequence $\{x_{n_i} : i \geq 1\}$ that is either increasing ($x_{n_i} \leq x_{n_{i+1}}$ for all $i \geq 1$) or decreasing ($x_{n_i} \geq x_{n_{i+1}}$ for all $i \geq 1$). (Why?) Suppose that $\{x_{n_i}\}$ is decreasing. By the Completeness Property, $\inf\{x_{n_i} : i \geq 1\}$ exists; denote it by x. Since the subsequence is decreasing, it must converge to x. (The reader should justify this.) $\square$

Exercises

(1) Let S be any nonempty set of real numbers. Show that a number α is the supremum of S if and only if α is an upper bound for S and for any $\varepsilon > 0$ there is $t \in S$ such that $t > \alpha - \varepsilon$.

(2) Justify all the steps in the proof of Theorem A.1.2.

(3) A sequence $\{x_n\}_{n>0}$ in $\mathbb{R}$ is said to be a **Cauchy sequence** if for all $\varepsilon > 0$ there exists an integer $N > 0$ such that whenever $m, n > N$ we have $|x_n - x_m| < \varepsilon$. Show that if $\{x_n\}_{n>0}$ is a Cauchy sequence, then there exists $x \in \mathbb{R}$ such that $\lim_{n \to \infty} x_n = x$.

(4) (Nested Intervals Property) Show that if $\{I_i\}_{i=1}^{\infty}$ is a countable collection of closed bounded intervals such that $I_{i+1} \subset I_i$, then $\bigcap_{i=1}^{\infty} I_i \neq \emptyset$. If, in addition, $\lim_{i \to \infty} |I_i| = 0$ (where $|I|$ denotes the length of an interval I), show that $\bigcap_{i=1}^{\infty} I_i$ consists of a single point. (Hint: Use Theorem A.1.2.)

(5) Let A be a set and let $\{B_\alpha\}$ be a collection of sets indexed by some set Γ. Show the following distributive properties.
   (a) $A \cap (\bigcup_{\alpha \in \Gamma} B_\alpha) = \bigcup_{\alpha \in \Gamma} (A \cap B_\alpha)$.
   (b) $A \cup (\bigcap_{\alpha \in \Gamma} B_\alpha) = \bigcap_{\alpha \in \Gamma} (A \cup B_\alpha)$.

(6) (*De Morgan's Laws*) Let Γ be a set and let $\{G_\alpha\}$ be a collection of sets in some set X indexed by Γ. Show that

$$\left(\bigcup_{\alpha \in \Gamma} G_\alpha\right)^c = \bigcap_{\alpha \in \Gamma} G_\alpha^c$$

and

$$\left(\bigcap_{\alpha \in \Gamma} G_\alpha\right)^c = \bigcup_{\alpha \in \Gamma} G_\alpha^c.$$

Appendix B

Topology of $\mathbb{R}$ and Metric Spaces

This section is devoted to the topological properties of the real numbers that are used in Chapter 2. It also introduces metric spaces and covers some topological notions on metric spaces at the same time as their counterparts in the real line.

A set G in $\mathbb{R}$ is said to be **open** if every $x \in G$ has an open ball centered at x that is contained in G; namely, for each $x \in G$ there is a number $\varepsilon = \varepsilon(x) > 0$ such that $B(x, \varepsilon) \subset G$. The empty set $\emptyset$ is seen to trivially satisfy the condition for being open (if it were not open there would exist an $x \in \emptyset$ for which there is no such an ε, but there is no x in $\emptyset$); one may also decide to just define the empty set to be open. $\mathbb{R}$ is also immediately seen to be open. We next observe that open intervals are indeed open sets. For if $I = (a, b)$ is a nonempty open interval and $x \in I$, one can take ε to be any positive number less than $\min\{b - x, x - a\}$; then clearly $B(x, \varepsilon) \subset (a, b)$. The following proposition lists the main properties of open sets.

Proposition B.1.1. *The open sets in $\mathbb{R}$ satisfy the following properties.*

(1) *The empty set $\emptyset$ and the set $\mathbb{R}$ are open.*

(2) *The union of any collection of open sets is open.*

(3) *The intersection of any finite collection of open sets is open.*

Proof. Part (1) is clear. To show (2), let $\{G_\alpha\}$ be a collection of open sets indexed by Γ, and let $G = \bigcup_{\alpha \in \Gamma} G_\alpha$. For any $x \in G$ there is some $\alpha \in \Gamma$ such that $x \in G_\alpha$. As G_α is open, there exists $\varepsilon > 0$ such that $B(x, \varepsilon) \subset G_\alpha$. Since $G_\alpha \subset G$, this shows that G is open.

(3) We show that the intersection of any two open sets is open; the general case follows by induction and is left to the reader. Let G_1 and G_2 be open sets and set $G = G_1 \cap G_2$. If G is empty, we are done; if not, let x be in G. As x is an element of G_1 and G_2 and these sets are open, for each $i = 1, 2$, there exist $\varepsilon_i > 0$ so that $B(x, \varepsilon_i) \subset G_i$. If $\varepsilon = \min\{\varepsilon_1, \varepsilon_2\}$, then $\varepsilon > 0$ and $B(x, \varepsilon) \subset G$, so G is open. □

It is important to remember that the intersection of an infinite collection of open sets need not be open. Indeed, for the open intervals $G_i = (-\frac{1}{i}, \frac{1}{i})$, $i \geq 1$, we have $\bigcap_{i=1}^{\infty} G_i = \{0\}$, and this set is not open.

The important concept to define the notion of open sets is that of an open ball, and this in turn is defined using the notion of distance. Now we generalize this notion of distance to introduce a very useful and general kind of space. A nonempty set X is said to be a **metric space** if there is a function $d : X \times X \to [0, \infty)$, called a **metric**, such that, for all points x, y, z in X,

(1) $d(x, y) = 0$ if and only if $x = y$;

(2) $d(x, y) = d(y, x)$;

(3) (Triangle Inequality) $d(x, y) \leq d(x, z) + d(z, y)$.

Examples. Euclidean distance provides us with an example of a metric. Let $X = \mathbb{R}$ and define d by $d(x, y) = |x - y|$ for $x, y \in \mathbb{R}$. There is a similar definition when $X = \mathbb{R}^d$: $d(x, y) = |x - y|$, where $|x - y| = \sqrt{(x_1 - y_1)^2 + \cdots + (x_d - y_d)^2}$. There is also a simple metric that can be defined on any set X by setting $d(x, y) = 1$ when $x \neq y$ and $d(x, x) = 0$ for all $x \in X$. Also, if (X, d) is a metric space, any subset A of X with d restricted to A provides an example of a metric space.

If (X, d) is a metric space, an **open ball** around $x \in X$ of radius $\varepsilon > 0$ is a set of the form $B(x, \varepsilon) = \{y \in X : d(x, y) < \varepsilon\}$. A set G in X is said to be **open** if for each $x \in G$ there is a number

B. Topology of the Real Numbers

$\varepsilon = \varepsilon(x) > 0$ such that $B(x, \varepsilon) \subset G$. The reader should note that formally the definition is the same as that for open subsets of $\mathbb{R}$. The basic properties also hold. The reader is asked to prove the following.

Proposition B.1.2. *Let (X, d) be a metric space. Then:*

(1) *The empty set $\emptyset$ and the set X are open.*

(2) *The union of any collection of open sets is open.*

(3) *The intersection of any finite collection of open sets is open.*

We give the following definitions in the context of metric spaces but discuss them in detail for $\mathbb{R}$. We assume that the reader is familiar with the notion of the limit of a sequence in $\mathbb{R}$ and state it for metric spaces for completeness. Let (X, d) be a metric space. Let $\{x_n\}$ be a sequence of points in X. We say that $\{x_n\}$ **converges** to a point $x \in X$ and write $\lim_{n \to \infty} x_n = x$ (or $x_n \to x$) if for all $\varepsilon > 0$ there exists an integer $N > 0$ such that $d(x_n, x) < \varepsilon$ for all $n > N$. Using this we can define the important notion of a dense set. A set $D \subset X$ is said to be **dense** in X if for each point $x \in X$ there exists a sequence $a_n \in D$ such at a_n converges to x. For example, the set $\mathbb{Q}$ is dense in $\mathbb{R}$.

If there exists a sequence $x_n \in A$ converging to x such that $x \neq x_n$ for all n, then x is called an **accumulation point** of the set A. A point $x \in X$ is an **isolated point** of a set $A \subset X$ if $x \in A$ and there is a number $\varepsilon > 0$ such that the open ball $B(x, \varepsilon)$ contains no points of A other than x. Thus, a point of A is an isolated point if and only if it is not an accumulation point. For example, if $A = \{1/n : n > 0\}$ in $\mathbb{R}$, then all its points are isolated points.

A set $F \subset X$ is **closed** if whenever $\{x_n\}$ is a sequence of points in F that converges to a point x, then $x \in F$. Since for every point of F we can take $x_n = x$, a set is closed if and only if it is equal to the union of its set of accumulation points with its set of isolated points. A set is **perfect** if it is equal to its set of accumulation points. One of the most interesting examples of a perfect set is the Cantor set studied in Section 2.2. Define the **closure** of a set A, denoted $\overline{A}$, to be the set of all points $x \in X$ such that there is a sequence $x_n \in A$ with x_n converging to x.

Exercise. Show that every perfect set is closed, and give an example of a closed set that is not perfect.

We next give a very useful characterization of closed sets.

Proposition B.1.3. *A set F is closed if and only if its complement F^c is open.*

Proof. First we show that if F^c is not open, then F is not closed. As F^c is not open, there exists a point $x \in F^c$ such that for each $n \geq 1$, if $\varepsilon_n = \frac{1}{n}$, the ball $B(x, \varepsilon_n)$ is not wholly contained in F^c, so there is a point $x_n \in F$ with $d(x, x_n) < \varepsilon_n$. This means that x_n converges to x and as $x \notin F$, then F is not closed.

For the converse suppose that F is not closed. We will show that F^c is not open. There is a sequence of points $x_n \in F$ whose limit x is not in F. Thus $x \in F^c$, but for any $\varepsilon > 0$ the ball $B(x, \varepsilon)$ contains points in the sequence. Therefore for all $\varepsilon > 0$, $B(x, \varepsilon)$ is not contained in F^c, so F^c is not open. □

The proof of the following proposition is left to the reader.

Proposition B.1.4. *Let (X, d) be a metric space. Then:*

(1) *The empty set $\emptyset$ and the set X are closed sets.*

(2) *The intersection of an arbitrary collection of closed sets is closed.*

(3) *The union of any finite collection of closed sets is closed.*

Theorem B.1.5 below is a special case of the important Heine–Borel Theorem; we follow Borel's original argument. For our development of Lebesgue measure on the real line in Chapter 2 we will need only the special case below; the general version appears in the exercises. A sequence of sets $\{A_n\}_{n=1}^{\infty}$ is said to **cover** a set A if $A \subset \bigcup_{n=1}^{\infty} A_n$.

Theorem B.1.5 (Heine–Borel—Special Case). *If $\{I_n\}_{n \geq 1}$ is a sequence of open intervals that covers a closed interval $I = [a, b]$, then there exists a finite subsequence of the intervals $\{I_{n_k}\}_{k=1}^{K}$ that covers I.*

B. Topology of the Real Numbers

Proof. Write $I_n = (a_n, b_n)$ for $n \geq 1$. Let n_1 be the smallest integer such that $a \in (a_{n_1}, b_{n_1})$. If $b < b_{n_1}$, then let $K = 1$ and we are done; otherwise there must exist a smallest integer such that $b_{n_1} \in (a_{n_2}, b_{n_2})$. If $b < b_{n_2}$ let $K = 2$; otherwise choose a smallest integer n_3 in a similar manner as before. We claim that this process stops for some integer n_K, in which case we are done with the proof. Now, if the process did not stop it would generate an infinite sequence $n_1, n_2, \ldots$, with $b_{n_k} < b_{n_{k+1}} \leq b$ for all $k \geq 1$. Then let $b^* = \lim_{k \to \infty} b_{n_k}$. As $\{b_{n_k}\}_{k \geq 1}$ is increasing, this limit exists and $b^* \leq b$. So there exists an interval I_p such that $b^* \in I_p$. Then $I_p = (a_p, b_p)$ is not one of the intervals I_{n_k} as $b_{n_k} \leq b^* < b_p$ for all $k \geq 1$; we shall see that this leads to a contradiction. Let ℓ be the first integer such that $n_{\ell+1} > p$ and $a_p < b_{n_\ell}$. This can be done since there are infinitely many integers n_k, the sequence b_{n_k} converges to b^* and $a_p < b^*$. Then $b_{n_\ell} \in I_p$, and this means that I_p should have been chosen after step n_ℓ instead of $I_{n_{\ell+1}}$, a contradiction. Therefore the process must terminate at some n_K. □

One may wonder why in Theorem B.1.5 we restrict ourselves to a cover consisting of countably many elements. However, it can be shown that since we are considering subsets of $\mathbb{R}$, any cover by an arbitrary family of open sets has a countable subcover (see Exercise 14). Therefore, in the case of $\mathbb{R}$ one does not obtain greater generality by considering arbitrary covers. We also note that the full version of the Heine–Borel theorem, which we do not need (see Exercise 12), applies to the case when I is an arbitrary closed bounded set and not just a closed bounded interval.

The following lemma will be used later and can be obtained as a consequence of the Bolzano–Weierstrass Theorem. The details of the proof are left as an exercise for the reader.

Lemma B.1.6. *If E and F are two disjoint closed bounded nonempty sets in $\mathbb{R}$, then there exists a $\delta > 0$ such that for any interval I, if $|I| < \delta$, then I cannot have a nonempty intersection with both E and F.*

Proof. Assume to the contrary that for any $\delta > 0$ there is an interval I with $|I| < \delta$ and such that $I \cap E \neq \emptyset$ and $I \cap F \neq \emptyset$. Then for all

$n \geq 1$, letting $\delta = 1/n$, we obtain points $x_n \in I \cap E$ and $y_n \in I \cap F$ such that $|x_n - y_n| < 1/n$. Since $\{x_n\}$ is bounded, by Theorem A.1.2 there exists a subsequence x_{n_i} such that $x_{n_i} \to x$ for some x in E. Observing that $|y_{n_i} - x| \leq |y_{n_i} - x_{n_i}| + |x_{n_i} - x|$, it follows that $y_{n_i} \to x$. But $x \in E$ and the points y_{n_i} are in F for all $i \geq 1$, a contradiction since F is closed. □

We conclude with a brief discussion of compactness in metric spaces. A metric space (X, d) is said to be **compact** if every sequence in X has a subsequence converging to a point in X. By Theorem A.1.2 a closed bounded interval in $\mathbb{R}$ is compact. A compact set K in X must be closed, as a converging sequence of elements of K must converge to a point of K. It must also be bounded, for otherwise one could construct a sequence diverging to ∞. For the converse we need another notion. A metric space (X, ε) is **totally bounded** if for all $\varepsilon > 0$ there exist finitely many point $x_1, \ldots, x_n$ such that for all $x \in X$ there exists $1 \leq i \leq n$ such that $x \in B(x_i, \varepsilon)$. Clearly, a totally bounded space is bounded. A sequence $\{x_n\}$ in X is a Cauchy sequence if for all $\varepsilon > 0$ there exists N such that $d(x_n, x_m) < \varepsilon$ for all $n, m > N$. We say that a metric space is **complete** if every Cauchy sequence in X converges to a point in X. If X is complete, then so is every closed subset of X. Exercise A.3 shows that $\mathbb{R}$ is complete. We state the following without proof (see, for example, [11, Theorem 9.58]).

Theorem B.1.7. *A metric space is compact if and only if it is complete and totally bounded.*

We define an important property that is enjoyed basically by all the metric spaces we study. A metric space (X, d) is said to be **separable** if it has a countable subset D that is dense in X. For example, $\mathbb{R}$ with the Euclidean metric is separable as $\mathbb{Q}$ is countable and dense in $\mathbb{R}$. It can be shown that a compact metric space is separable (Exercise 13).

A Special Subset of $\mathbb{R}$

We construct a subset K^* of the reals based on the Cantor set that can be used to construct various counterexamples.

B. Topology of the Real Numbers

First let K be a Cantor set of measure $1/2$ in $[0,1]$ (see, e.g., Exercise 2.4.5). Now for each open interval I in $[0,1]$ we define a modified Cantor-like set $K(I)$ as follows. First delete the endpoints of K and denote this new set by K', i.e., let $K' = K \setminus \{0,1\}$. Then we obtain a new copy of K' by scaling K' by the factor $\lambda(I)$ and placing the scaled copy at the left endpoint of I. This new set is denoted by $K(I)$. This set maintains the properties of the Cantor set that we need below.

The construction of the set K^* is done by induction. First define

$$L_1 = K((0,1/2)) \text{ and } M_1 = K((1/2,1)).$$

For concreteness we define L_2 and M_2. Note that the complement of L_1 in $(0,1/2)$ consists of the union of countably many disjoint open intervals $A_{1,i}$, and the complement of M_1 in $(1/2,1)$ consists of the union of countably many disjoint open intervals $B_{1,i}$. Set $M_2 = \bigcup_{i=1}^{\infty} K(A_{1,i})$ and $L_2 = \bigcup_{i=1}^{\infty} K(B_{1,i})$.

Assume now that L_n and M_n have been defined. If n is odd, then the complement of L_n in $(0,1/2)$ is a union of countably many disjoint open intervals $A_{n,i}$. Then set $M_{n+1} = \bigcup_{i=1}^{\infty} K(A_{n,i})$ and $L_{n+1} = \bigcup_{i=1}^{\infty} K(B_{n,i})$. When n is even, the complement of M_n in $(1/2,1)$ is the union of countably many disjoint open intervals $B_{n,i}$. Then set $L_{n+1} = \bigcup_{i=1}^{\infty} K(A_{n,i})$ and $M_{n+1} = \bigcup_{i=1}^{\infty} K(B_{n,i})$.

We are ready to define K^* by

$$K^* = \bigcup_{n=1}^{\infty} L_n.$$

We observe that $\bigcup_{n=1}^{\infty} M_n \subseteq K^{*c}$. The following proposition shows the main property of this set.

Proposition B.1.8. *For any interval I of positive length in $[0,1]$,*

$$\lambda(I \cap K^*) > 0 \text{ and } \lambda(I \cap K^{*c}) > 0.$$

Proof. It suffices to show that for any open interval $I = (a,b)$ ($a < b$) in $[0,1]$, if $\lambda(I \cap K^*) > 0$, then $\lambda(I \cap K^{*c}) > 0$, and if $\lambda(I \cap K^{*c}) > 0$, then $\lambda(I \cap K^*) > 0$. So suppose that $\lambda(K^* \cap I) > 0$ and let $x \in K^* \cap I$. As every element of K^* is in some $K(A)$ for some open interval A, $x \in I \cap K(A)$. Moreover, by the construction of

$K(A)$, for $\varepsilon = \min\{x - a, b - x\} > 0$, there exists $y \in K(A)$ such that $|x - y| < \varepsilon$. Thus $y \in I$. As x and y are elements of the same set $K(A)$ and $K(A)$ is a totally disconnected set, there must exist some open interval contained in $(\min\{x, y\}, \max\{x, y\})$ that is part of $K(A)^c$. In the construction of K^*, we know that at the $(n+1)^{\text{st}}$ stage, a set of positive measure $K(B)$ was put in that open interval, and $K(B) \subset M_{n+1} \subset K^{*c}$. Then

$$\lambda(K^{*c} \cap I) \geq \lambda(K(B) \cap I) = \lambda(K(B)) > 0.$$

Next assume that $\lambda(I \cap K^{*c}) > 0$. We observe that any element $z \in K^*$ corresponds to an element $(1-z) \in K^{*c}$. The reader can verify that $\lambda(\bigcup_{n=1}^{\infty} L_n \cup \bigcup_{n=1}^{\infty} M_n) = 1$. This implies that the points in K^{*c} that do not correspond to elements in K^* by the symmetry above form a set of measure zero. Therefore, any interval $(1-d, 1-c)$ that intersects K^{*c} in positive measure and K^* in zero measure would correspond to an interval (c, d) which would intersect K^* in positive measure and K^{*c} in zero measure, which is a contradiction. □

Exercises

(1) Show that closed intervals are closed sets.

(2) Prove Proposition B.1.2.

(3) A point x is said to be an **interior point** of a set A if there is a number $\varepsilon > 0$ such that the open ball $B(x, \varepsilon)$ is in A. Let $\text{Int}(A)$ denote the set of interior points of A. a) Show that for any set A, $\text{Int}(A)$ is open. b) Show that for any set A, A is open if and only if $\text{Int}(A) = A$.

(4) a) Show that for any set A, $\text{cl}(A)$ is closed. b) Show that for any set A, A is closed if and only if $\text{cl}(A) = A$.

(5) Is there a set whose interior is empty and whose closure is $\mathbb{R}$?

(6) Complete the proof of Proposition B.1.4.

(7) Find a countably infinite collection of closed intervals that cover $[0, 1]$ but such that no finite subcollection covers $[0, 1]$.

(8) Show that a perfect set in $\mathbb{R}$ must be uncountable.

B. Topology of the Real Numbers

* (9) Show that any nonempty open subset G of $\mathbb{R}$ can be written as a countable union of disjoint finite or infinite open intervals. (Hint: For each x in G show that there exists a "largest" interval I_x with $x \in I_x \subset G$.)

(10) Two intervals are said to be nonoverlapping if they intersect in at most one point. Show that any nonempty open set can be written as a countable union of nonoverlapping closed finite intervals.

(11) Justify all the steps in the proof of Lemma B.1.6.

* (12) (Heine–Borel Theorem) Show that if $\{G_n\}_{n\geq 1}$ is a sequence of open sets that covers a closed bounded set $A \subset \mathbb{R}$, then there exists a finite subcollection $\{A_{n_k}\}_{k=1}^{K}$ that covers A. Show that the converse is also true. Deduce that a set $A \subset \mathbb{R}$ is closed and bounded if and only if every open cover of A has a finite subcover.

(13) Show that a compact metric space is separable.

* (14) (Lindelöf) Show that if $\{G_\alpha\}_{\alpha \in \Gamma}$ is an arbitrary collection of open sets in $\mathbb{R}$, then there exists a countable subcollection G_i such that $\bigcup_{\alpha \in \Gamma} G_\alpha = \bigcup_{i=1}^{\infty} G_i$. (Hint: Use open intervals with rational endpoints.)

Bibliographical Notes

Chapter 2

For further information about these topics the reader is referred to [**67**] and [**56**].

Our definition of Lebesgue measurable set is as in [**70**] and our presentation follows [**56**] and [**70**]. Our definition of semi-rings is as in [**68**]. For the sections on σ-algebras, measure spaces and the monotone class theorem we have followed [**28**] and [**68**]. For Section 2.8 we have followed [**68**], [**6**], [**18**] and [**28**]. The quotation in Section 2.7 on Littlewood's Three Principles is from [**61**]. For a short account of the history of measure and integration with extensive references see [**18**].

Chapter 3

For further information about these topics the reader is referred to [**24**], [**58**] and [**69**]. An introduction to several topics in dynamics can be found in [**31**] and then in [**10**] (including an application of ergodic theory to Internet searches). For entropy theory, which is not covered here, the reader may start with the entropy chapter in [**50**] and then consult [**39**], [**58**] and [**69**].

The baker's transformation is a well-known example in ergodic theory. For a higher-dimensional version of irrational rotations see

[15] and [57]. For the topological dynamics examples we have followed [69]. For applications of the Baire Category method to dynamics see [3] and [56]. For a discussion of nonmeasurable sets see [56]. The proof of the Baire category theorem is standard (see, for example, [56]). For measure-preseving transformations and recurrence we have followed [24], [58] and [69]. For multiple recurrence refer to [24]. The notions of Poincaré sequence and thick sets are from [24]. For the recent proof of Green and Tao on arithmetic progressions in the primes and the role of Furstenberg's ergodic theoretic methods in this proof see [43]. The notion of ergodicity goes back to a paper of Birkhoff and Smith [7]. Most of the equivalences of Lemma 3.7.2 are in, for example, [58]. Part (4) of Lemma 3.7.2 is probably known but I learned it from Daniel Kane, as well as Exercise 6.2.5. The dyadic odometer is due to Kakutani and von Neumann; a description similar to ours can be found in [21]. The Hajian-Kakutani transformation is from [27], and weakly wandering sets were introduced in [26]. For further results on infinite measure-preserving transformations see [1]. For factors and isomorphism we have followed [24] and [63]. The notion of a Lebesgue space is due to Rohlin; for properties of Lebesgue spaces see [48] and [63]. The induced transformation is due to Kakutani [37]. For further properties of symbolic systems refer to [48] and [46]. Furstenberg's question may be found in [64].

Chapter 4

Our development of Lebesgue integration follows [11]. For the Gauss transformation see [48]. Other examples of invariant measures may be found in [8]. For other number theoretic examples see [16]. For the Lebesgue L^p spaces we have followed [11] and [70]. The results on eigenvalues are standard; see, for example, [69]. For product measure we have followed [6] and [68].

Chapter 5

The Birkhoff ergodic theorem has had a long history of different proofs. For our first proof of the ergodic theorem we have followed [66], which was influenced by [54] and [35]; these arguments can be extended to the ratio subadditive ergodic theorem [66]. Our second proof of the ergodic theorem follows [29]. For another proof of the maximal ergodic theorem due to Garsia see, e.g., [**57**]. A more recent proof of the ergodic theorem with references to other proofs is in [**41**]. For a comprehensive survey of developments related to the ergodic theorem see [**45**] and [**60**]. For the proof of the L^2 ergodic theorem we followed [**57**]. The proof of Lemma 5.4.1 is from [**67**]. A proof of Theorem 5.4.5 can be found in [**69**]. The proof of Lemma 5.4.3 is from [**71**]. For additional material on equidistribution see [**57**]. For other expositions of the ergodic theorem see [**39**], [**58**], [**65**], [**69**].

Chapter 6

Many of the characterizations of weak mixing go back to Koopman and von Neumann and already appear in [**29**]. We have followed [**19**], [**24**], [**58**], [**69**]. The notion of double ergodicity and its equivalence to weak mixing appears in [**24**]; it was generalized to infinite measure-preserving transformations in [**9**]. For a proof of the existence of sequences of density one for a weakly mixing transformation see [**20**], [**69**] and [**58**]. For Section 6.4 we have followed [**58**] and [**63**]. Chacón's transformation appears in [**19**], and appeared in a modified form in [**13**]. The transformation in [**13**] is lightly mixing (hence weakly mixing) but not mixing [**22**]. The transformation in [**19**] is weakly mixing (and mildly mixing [**22**]) but not lightly mixing [**19**], [**22**]. The proof of Theorem 6.5.2 follows [**13**] and [**19**]. The proof of Lemma 6.5.4 (double approximation) follows a proof in [**2**] and [**17**]. The proof of Theorem 6.5.5 follows a proof in [**2**] of a similar result in infinite measure. Another proof of weak mixing is in [**63**]. Other examples of weakly mixing and not mixing transformations are in [**40**] and [**38**]. The canonical Chacón transformation was shown to be prime (no nontrivial factors) and to commute only with its powers in [**36**]. The notion of mild mixing and rigidity is from [**25**]. For the proof of Lemma 6.7.3 we followed [**2**], where it is in the more general context of nonsingular transformations. For the proof of Proposition 6.7.5 we followed [**32**]. Section 6.8 is from [**51**]. Exercise 6.8.5 is from [**47**].

Appendix A

For basic properties of sets see [**44**]. For an introduction to set theory see [**30**]. For properties of the real numbers the reader may consult [**44**], [**49**], [**62**]. For measurable dynamics on other metric completions of the field $\mathbb{Q}$ see [**12**] and [**42**].

Appendix B

The proof of the Heine–Borel Theorem (Theorem B.1.5) follows Borel's proof as in [**56**, p. 4]. For the set K^* we followed [**51**] and an early version of [**51**]. For the other topics see [**44**], [**49**], [**62**].

Bibliography

[1] J. Aaronson, *An Introduction to Infinite Ergodic Theory*, American Mathematical Society, 1997.

[2] T. Adams, N. Friedman, and C.E. Silva. Rank-one weak mixing for nonsingular transformations, *Israel J. Math.* **102** (1997), 269–281.

[3] S. Alpern and V.S. Prasad, *Typical dynamics of volume preserving homeomorphisms*, Cambridge, 2000.

[4] V.I. Arnold and A. Avez, *Ergodic Problems of Classical Mechanics*, W.A. Benjamin, 1968.

[5] P. Billingsley, *Ergodic Theory and Information*, Wiley, 1965.

[6] P. Billingsley, *Probability and Measure*, Wiley, 1986.

[7] G.D. Birkhoff, *Collected Mathematical Papers*, 3 vols., American Mathematical Society, New York, 1950.

[8] A. Boyarski and P. Gora, *Laws of Chaos: Invariant Measures and Dynamical Systems in One Dimension*, Birkhäuser, 1997.

[9] A. Bowles, L. Fidkowski, A. Marinello, and C.E. Silva, Double ergodicity of nonsingular transformations and infinite measure-preserving staircase transformations, *Illinois J. Math.* **45** (2001), no. 3, 999–1019.

[10] M. Brin and G. Stuck, *Introduction to Dynamical Systems*, Cambridge, 2002.

[11] A. Bruckner, J. Bruckner, and B. Thomson, *Real Analysis*, Prentice Hall, 1997.

[12] J. Bryk and C.E. Silva, Measurable dynamics of simple p-adic polynomials, *Amer. Math. Monthly* **112** (2005), no. 3, 212–232.

[13] R.V. Chacón, Weakly mixing transformations which are not strongly mixing, *Proc. Am. Math. Soc.* **22** (1969), 559–562.

[14] D.C. Cohn, *Measure Theory*, Birkhäuser, 1980.

[15] I.P. Cornfeld, S.V. Fomin, and Ya. G. Sinai, *Ergodic Theory*, Springer-Verlag, 1982.

[16] K. Dajani and C. Kraaikamp, *Ergodic Theory of Numbers*, Mathematical Association of America, 2002.

[17] S. Day, B. Grivna, E. McCartney, and C.E. Silva, Power weakly mixing infinite transformations, *New York J. Math.* **5** (1999), 17–24.

[18] R.M. Dudley, *Real Analysis and Probability*, Wadsworth & Brooks/Cole, 1989.

[19] N. Friedman, *Introduction to Ergodic Theory*, Van Nostrand Reinhold, 1970.

[20] N. Friedman, Mixing on sequences, *Canad. J. Math.* **35** (1983), no. 2, 339–352.

[21] N. Friedman, Replication and stacking in ergodic theory, *Amer. Math. Monthly* **99** (1992), no. 1, 31–41.

[22] N. Friedman and J.L. King, Rank one lightly mixing, *Israel J. of Math.* **73** (1991), no. 3, 281–288.

[23] H. Furstenberg, Strict ergodicity and transformation of the torus, *Amer. J. Math.* **83** (1961), 573–601.

[24] H. Furstenberg, *Recurrence in Ergodic Theory and Combinatorial Number Theory*, Princeton Univ. Press, 1981.

[25] H. Furstenberg and B. Weiss, The finite multipliers of infinite ergodic transformations, *The Structure of Attractors in Dynamical Systems*, Lecture Notes in Mathematics, Vol. 668, Springer, Berlin, 1978, 127–132.

[26] A. Hajian and S. Kakutani, Weakly wandering sets and invariant measures, *Trans. Amer. Math. Soc.* **110** (1964), 136–151.

[27] A. Hajian and S. Kakutani, An example of an ergodic measure, preserving transformation defined on an infinite measure space, *Contributions to Ergodic Theory and Probability*, Lecture Notes in Mathematics, vol. 160, Springer, Berlin, 1970, 45–52.

[28] P.R. Halmos, *Measure Theory*, Van Nostrand, 1950.

[29] P.R. Halmos, *Lectures on Ergodic Theory*, Chelsea, 1956.

[30] P.R. Halmos, *Naive Set Theory*, Van Nostrand, 1960.

[31] B. Hasselblatt and A. Katok, *A First Course in Dynamics*, Cambridge, 2003.

Bibliography

[32] J. Hawkins and C.E. Silva, Characterizing mildly mixing actions by orbit equivalence of products, *New York J. Math.* **3A** (1997/98), Proceedings of the New York Journal of Mathematics Conference, June 9–13, (1997), 99–115.

[33] E. Hopf, *Ergodentheorie*, Verlag, 1937.

[34] F. Jones, *Lebesgue Integration on Euclidean Space*, Jones and Bartlett, 1993.

[35] R.L. Jones, New proofs for the maximal ergodic theorem and the Hardy-Littlewood maximal theorem, *Proc. Amer. Math. Soc.* **87** (1983), no. 4, 681–684.

[36] A. del Junco, A simple measure-preserving transformation with trivial centralizer, *Pacific J. Math.* **79** (1978), 357–362.

[37] S. Kakutani, Induced measure-preserving transformations, *Proc. Japan Acad.* **19** (1943), 635–641.

[38] S. Kakutani, Examples of ergodic measure preserving transformations which are weakly mixing but not strongly mixing, *Recent advances in topological dynamics*, Lecture Notes in Mathematics, Vol. 318, Springer, Berlin, 1973, 143–149.

[39] A. Katok and B. Hasselblatt, *Introduction to the Modern Theory of Dynamical Systems*, Cambridge, 1995.

[40] A.B. Katok and A.M. Stepin, Approximation in ergodic theory, *Russian Mathematical Surveys* **22** (1967), 77–102.

[41] M. Keane and K. Petersen, *Easy and Nearly Simultaneous Proofs of the Ergodic Theorem and Maximal Ergodic Theorem*, IMS Lecture Notes–Monograph Series, Vol. 48, Inst. Math. Stat., 2006, 248-251.

[42] A.Y. Khrennikov and M. Nilson, *P-adic deterministic and random dynamics*, Kluwer Academic Publishers, Dordrecht, 2004.

[43] B. Kra, The Green-Tao theorem on arithmetic progressions in the primes: an ergodic point of view, *Bulletin Amer. Math. Soc. (N. S.)* **43** (2006), 3–23.

[44] S. Krantz, *Real Analysis and Foundations*, CRC Press, 1991.

[45] U. Krengel, *Ergodic Theorems*, de Gruyter Studies in Mathematics, Vol. 6, Walter de Gruyter & Co., Berlin-New York, 1985.

[46] D. Lind and B. Marcus, *An introduction to Symbolic Dynamics and Coding*, Cambridge University Press, 1995.

[47] D. Maharam, On orbits under ergodic measure-preserving transformations, *Trans. Amer. Math. Soc.* **119** (1965), 51–66.

[48] R. Mañé, *Ergodic Theory and Differentiable Dynamics*, Springer-Verlag, 1987.

[49] F. Morgan, *Real Analysis,* Amer. Math. Soc., 2005.

[50] D. Witte Morris, *Ratner's Theorems on Unipotent Flows,* Univ. of Chicago Press, 2005.

[51] E. Muehlegger, A. Raich, C.E. Silva, and W. Zhao, Lightly mixing on dense algebras, *Real Anal. Exchange* **23** (1997/8), 259–266.

[52] M.G. Nadkarni, *Basic Ergodic Theory,* Second Edition, Birkh"auser, 1995.

[53] D. Ornstein, *Ergodic Theory, Randomness and Dynamical Systems,* Yale Univ. Press, 1974.

[54] D. Ornstein and B. Weiss, The Shannon-McMillan-Breiman theorem for a class of amenable groups, *Israel J. Math.* **44** (1983), no. 1, 53–60.

[55] J. Oxtoby, Ergodic sets, *Bull. Amer. Math. Soc.* **58** (1952), 116–136.

[56] J. Oxtoby, *Measure and Category,* Second Edition, Springer-Verlag, 1980.

[57] W. Parry, *Topics in Ergodic Theory,* Cambridge, 1981.

[58] K. Petersen, *Ergodic Theory,* Cambridge, 1983.

[59] V.A. Rokhlin, On the fundamental ideas of measure theory, *Mat. Sb.* **25**, 107-50. Amer. Math. Soc. Transl. **71**, 1952.

[60] J.M. Rosenblatt and M. Wierdl, Pointwise ergodic theorems via harmonic analysis, *Ergodic theory and its connections with harmonic analysis (Alexandria, 1993)*, London Math. Soc. Lecture Note Ser., 205, 3–151, Cambridge, 1995.

[61] H.L. Royden, *Real Analysis,* Macmillan, 1988.

[62] W. Rudin, *Principles of Mathematical Analysis,* MacGraw-Hill, 1976.

[63] D. Rudolph, *Fundamentals of Measurable Dynamics,* Oxford, 1990.

[64] D. Rudolph, ×2 and ×3 invariant measures and entropy, *Ergodic Theory Dynam. Systems* **10** (1990), no. 2, 395–406.

[65] P.C. Shields, *The Ergodic Theory of Discrete Sample Paths,* Amer. Math. Soc., 1996.

[66] C.E. Silva and P. Thieullen, The subadditive ergodic theorem and recurrence properties of Markovian transformations, *J. Math. Anal. Appl.* **154** (1991), no. 1, 83–99.

[67] E.M. Stein and R. Shakarchi, *Real Analysis,* Princeton, 2005.

[68] S.J. Taylor, *Introduction to Measure and Integration,* Cambridge, 1966.

[69] P. Walters, *An Introduction to Ergodic Theory,* Springer-Verlag, 1981.

[70] R. Wheeden and A. Zygmund, *Measure and Integral: An Introduction to Real Analysis,* Dekker, New York-Basel, 1977.

[71] D. Williams, *Probability with Martingales,* Cambridge, 2004.

Index

$(1-\delta)$-full, 98
L^2 inner product, 162
L^∞-norm, 162
L^p-norm, 193
T-invariant, 91
T-invariant mod μ, 92
σ-algebra, 26, 27
σ-algebra generated by, 34
σ-finite, 29
σ-ideal, 82
d-dimensional Lebesgue outer measure, 51
d-volume, 51
p-norm, 161, 193
r_n-odometer, 108

a.e., 91
absolutely continuous, 151
absolutely normal, 190
accumulation point, 243
algebra, 36
almost disjoint, 25
almost everywhere, 91
atom, 30

Baire Category Theorem, 79
Bernstein set, 75
Boole's transformation, 85
Borel measurable, 126
Borel measurable function, 134

Borel sets, 35
Borel–Cantelli, 33
bounded above, 238
bounded below, 237
bounded interval, 237

canonical atomic spaces, 29
canonical Chacón transformation, 219
canonical Lebesgue measure space, 30
canonical nonatomic Lebesgue measure space, 29
canonical representation of a simple function, 142
Cantor middle-thirds set, 10
Cantor set, 14
Cartesian product, 51
Cauchy sequence, 239
Cesàro convergence of sequences, 202
characteristic function, 77
closed set, 243
closure, 243
column, 104
compact, 246
complete, 246
complete measure space, 29
compressible, 88
conjugate, 160

259

conservative, 87
continued fraction expansion, 154
continuous, 71
continuous spectrum, 169
converge in density, 204
convergence of sequences, 239
converges, 243
copy, 220
countable basis, 166
countably subadditive, 48
counting measure, 30
cutting and stacking, 114

dense, 243
dense algebra, 44
dense ring, 44
density one, 203
doubling map, 76
doubly ergodic, 207
dyadic interval, 9
dynamical property, 118

eigenfunction, 167
eigenvalue, 166
eigenvalue group, 168
eigenvector, 167
element, 235
empty set, 235
equal almost everywhere, 137
equivalent mod 1, 67
equivariance, 118
ergodic, 96
essential supremum, 162
exhaustive, 112

factor, 119
fiber, 172
finite measure space, 29
finite measure-preserving, 83
finitely additive, 47
first category, 78
first return time , 121
full measure, 116

Gauss map, 152
Gelfand's question, 72
generate mod 0, 44

height, 104

homeomorphism, 130

improper σ-algebra, 27
incompressible, 88
indicator function, 77
induced transformation, 121
infimum, 238
infinite measure-preserving, 83
integrable, 156
interior point, 248
interval, 237
invariant, 91
invariant measure, 69
inverse image, 69
invertible measurable
 transformation, 69
invertible measurable
 transformation mod μ, 94
invertible measure-preserving, 70
invertible transformation, 67
isolated point, 243
isomorphic, 116, 118
isomorphism, 118

least period, 68
Lebesgue integrable, 145, 156
Lebesgue integral of a
 characteristic function, 141
Lebesgue integral of a measurable
 function, 156
Lebesgue integral of a nonnegative
 function, 145
Lebesgue integral of a simple
 function, 142
Lebesgue measurable, 17
Lebesgue measurable function, 134
Lebesgue measure, 23
Lebesgue outer measure, 6
Lebesgue space, 117
length, 237
level, 104
lightly mixing, 225
Liouville number, 16
lower bound, 237

meager, 78
mean convergence, 193
measurable function, 134
measurable rectangle, 171

Index

measurable transformation, 69
measure, 28
measure space, 29
measure-preserving, 69
measure-preserving dynamical
    system, 83, 118
measure-preserving isomorphism
    mod 0, 116
mesh, 131
metric, 242
metric space, 242
mildly mixing, 229
minimal, 72
mixing sequence, 214
monotone class, 36
monotone class generated by, 36
monotone set function, 48
multiply recurrent, 89

negatively nonsingular, 152
nonsingular, 152
norm, 161
normal to base 2, 189
normed linear space, 162
nowhere dense, 14, 78
null set, 9

odometer map, 127
open, 242
open ball, 242
open bounded interval, 237
open set, 241
orbit, full, 68
orbit, positive, 68
orthogonal collection, 164
orthogonal complement, 194
orthogonal functions, 164
orthonormal collection, 164
outer measure, 48

pairwise disjoint, 236
partially rigid, 228
partition, 131
perfect set, 243
period, 68
periodic, 68
periodic point, 68
Poincaré Recurrence Theorem, 88
Poincaré sequence, 90

positive density, 203
positively invariant, 91
power set, 27
pre-image, 69
probability space, 29
probability-preserving, 83
product measure, 171
projection, 195

rational eigenvalue, 168
real and imaginary part of a
    function, 160
recurrent, 86
residual, 82
Riemann integrable, 132
rigid, 227
ring, 44
ring generated by, 46
rotation by α, 67

semi-ring, 39
separable metric space, 246
set function, 47
sets restricted to Y, 28
shift, 127
simple eigenvalue, 167
simple function, 141
simply normal number, 77
simply normal to base 2, 188
spacers, 110, 114
square integrable, 160
strictly invariant mod μ, 92
strictly invariant, 91
strictly periodic, 68
strong Cesàro convergence of
    sequences, 202
sublevels, 220
subset, 235
sufficient ring, 209
sufficient semi-ring, 41
supremum, 238
sweeps out, 97
symbolic binary representation, 76
symmetric difference, 235

tent map, 78
thick set, 90
topologically transitive, 78
totally disconnected, 14

totally ergodic, 101
tower, 104
transformation, 67
triadic interval of order n, 222
trivial σ-algebra, 27

unbounded interval, 237
uniformly distributed, 191
upper bound, 237

wandering, 90, 112
weakly mixing, 205
weakly wandering, 112

zero density, 203